#DoNotDisturb

#DoNotDisturb

How I Ghosted
My Cell Phone
to Take Back
My Life

Jedediah Bila

HARPER

An Imprint of HarperCollins*Publishers*

HarperCollins books may be purchased for educational, business, or sales promotional use. For information, please email the Special Markets Department at SPsales@harpercollins.com.

FIRST EDITION

DESIGNED BY WILLIAM RUOTO

Library of Congress Cataloging-in-Publication Data
Names: Bila, Jedediah, author.
Title: #DoNotDisturb : how I ghosted my cell phone to take back my life / Jedediah Bila.
Description: First edition. | New York, NY : Harper, [2018] | Includes bibliographical references.
Identifiers: LCCN 2018012567 (print) | LCCN 2018014849 (ebook) | ISBN 9780062797070 (ebk) | ISBN 9780062797063 (hardback)
Subjects: LCSH: Bila, Jedediah—Mental health. | Cell phones—Psychological aspects—Case studies. | Internet addiction—Case studies. | Television—United States—Biography. | BISAC: SOCIAL SCIENCE / Popular Culture. | BIOGRAPHY & AUTOBIOGRAPHY / Personal Memoirs.
Classification: LCC RC569.5.I54 (ebook) | LCC RC569.5.I54 B55 2018 (print) | DDC 616.85/84—dc23
LC record available at https://lccn.loc.gov/2018012567

18 19 20 21 22 LSC 10 9 8 7 6 5 4 3 2 1

CONTENTS

INTRODUCTION: PAGING MARTY MCFLY

I'VE BEEN BEGGING FOR A DELOREAN FOR MY BIRTHDAY for as long as I can remember. Every year I wake up, look out my window, and hope to see the poufy white hair and electrified eyes of Dr. Emmett Brown, his long arms outstretched from his too-short lab coat, waving excitedly, beckoning me forth, eager to send me back in time.

Back to the Future is one of my favorite movies. I've seen it at least twenty times. Mostly because I suffer from a particular condition called *Simple Life Envy*. Instead of marching out of my apartment into the hustle and bustle and technology-muscle that is our modern world, brushing shoulders with people scurrying here and there on their smartphones, yapping into wireless headphones, scrolling through the latest and greatest social media apps, I crave something different. I want to walk outside into a stream of people aware of their surroundings, looking at the detailed façade of an old brownstone, pointing out a rabbit-shaped cloud in the sky, connecting with the loved one beside them, taking hold of a friend's arm as they laugh uncontrollably at a joke well told. I want to see children playing with other children on playgrounds, no iPads or other such screens in sight, and their parents actually watching them, not looking down at a device in their hands or interacting with someone else who is somewhere else doing something else. I want to pass by people who say hello. I'd say hello back, then breeze along to the neighborhood diner, squeeze into a booth with friends who are deep in conversation, order breakfast for dinner, and appreciate the fact that the most high-tech item in our vicinity is a jukebox playing a great song from the 1950s or '60s. Perhaps the Shirelles, asking wistfully: "Will you still love me tomorrow?" Yep, that sounds about right.

It isn't pure fantasy to imagine myself in an era when people engaged meaningfully with one another, when outside distractions were few, when you could close chapters of your life and move on without an iCloud or social media reminder trail, when people looked into each other's eyes on first dates or glanced at a patch of red roses on a walk home.

Those days really did exist. I remember them. And it wasn't too long ago.

I recall a place where we were in that place and only that place, and only those we could see in front of us were there with us; a time when the frequent, nearby buzzing of a tiny nuisance demanding attention was most likely an insect, not a device for which we'd pay dearly every month; a moment when we were actually in the moment, and not taking a selfie of it.

For those of you who can't imagine such a time, believe me when I, and others who knew the world before the ubiquity of smartphones, tell you about it. It existed. And it was wonderful.

Back then, before these digital doohickeys dominated our world, we lived the lives we were living, instead of constantly trying to capture a perfect representation of those lives to post on social media, for us to then check obsessively for views. Or likes. Or whatever. Over and over.

While I know we can't go backward, I'd sure like to move forward in a better way. With an appreciation for the simple and the good, and a thoughtful intention to incorporate the best of what was, into what is and can be.

That said, I'm not one of those I-had-to-walk-five-miles-to-school-barefoot-in-the-snow-uphill militant memorialists, resentful of the ease that has come with progress. I like aspects of my smartphone and the laptop on which I typed this. I have a social media presence. I don't want to lose any of the wonderful ways the latest and greatest innovations have improved lives. Nor am I denying the ways in which technology has helped humankind in so many arenas, such as science, research, exploration, and communications. But I am beyond frustrated with the surge in personal technology that has compromised so many aspects of who we are and how we live. It's affecting all of us. Even if you don't know it.

If you have no idea of the extent to which tech is dominating your life, this book is for you.

If you have some idea, but no clue as to how to deal with it, this book is for you.

If you have some idea, but have resigned to let tech take over your brain, this book is also for you. Because your brain is a good thing, and you need to know and understand exactly what that gadget in your hand is doing to it.

How these gadgets that now rest constantly in our pockets, on our laps, and in our palms affect our hearts and minds, and our too-often blind acceptance of some new, brain-shrinking, behavior-altering status quo—that's the problem.

The solution? Well, I think about the tech issue in the same way that I think about many others that affect society. I believe that *we, the people*, are responsible for ourselves and our actions. When it comes to tech, we must ask questions and consider usage implications for our daily lives and relationships, before embracing Silicon Valley's latest hot trend. We must consider how we want to use technology so that we can design and build the lives we want, the lives that make us happy. What that looks like will be different for each person, and that's okay. The key is to remember that no matter what gadget or app or seemingly convenient who-knows-what emerges next, *we, the people*, hold the power to use it as we see fit. Or not at all.

That is why I went on this journey of exploration and examination to discover where I was and who I wanted to be, with and without my cell phone, computer, and social media. On this road, I asked myself twelve questions about what I felt I had lost in my tech-dominated life, and along the way I uncovered some answers, and even some solutions.

It wasn't easy to admit the things I reveal here, the details of how I lost myself in the thickets of the tech jungle. Some of these personal stories, reflections, discoveries, and slipups are embarrassing, even downright humiliating, but I'm sharing it all because I want you to know that there is a person out there who can help you disentangle from the web of technology that is overtaking your life and your sanity. That person is:

You.

I found my way out of the tech jungle by absorbing wisdom and advice from trusted resources, and then relying on the one person who could act to fix me:

Me.

Now, I don't suggest that you move to a remote town, toss your phone in a well, communicate by telegraph, and carry a lantern to and from your evening destinations. But I'm also not suggesting that you passively allow whatever intrusions some random software developer cooks up to enter your life unchallenged and unquestioned. Instead, I share the powerful package of homemade solutions I created for myself—an embrace of the tech features that enhance my life, with self-imposed limits that make me healthier, happier, and in charge.

My hope is that you will find yourself in the stories I share and come away empowered to own the tech in your life instead of letting the technology own you. If you want to. I hope you do, because the quality of our relationships and our experiences with each other, with nature and our surroundings, is at stake. The quality of life for our kids and grandkids depends upon us taking control. The next generation is going to miss so much of the beauty we grew up with if we don't stand tall in this fight and tell technology to sit the hell down.

This is our time to take action, to take control, and to take back our lives.

I'm hoping you'll put your phone away, or at least put it on airplane mode, and venture with me through these pages, where I examine the current cultural climate, consider the best and worst of what technology has to offer, and reveal the nitty-gritty, personal anecdotes from my voyage through the wringer of the tech revolution, as well as my attempts to navigate my way back out again, to *go back to the future* that is mine, and reclaim my life.

Maybe where we're going we don't need roads, but I sure as heck needed a little direction. I thought you might, too. So here we go . . .

## CHAPTER 1

MÉNAGE-À-TECH

Can I find authentic interaction in an
online world?

IT'S A STRANGE FEELING WHEN YOU FIND YOURSELF BLISS-
fully swept up in a fairy tale you never even knew you wanted.
That's how I felt the day of my wedding in February of 2018. For
years I had joked with my mom that I'd likely get married in my
denim shorts and combat boots, under a small canopy we'd stick in
the sand close to the sound of ocean waves. Then my friend Mikie
suggested that my fiancé, Jeremy, and I "just take a look" at Oheka
Castle, a French-style château in the quaint hamlet of Cold Spring
Harbor, New York.

"A *castle*?" I said in disbelief. "I don't know if that's for me."
Though I had watched Disney's *Cinderella* enough times to be intrigued by the castle concept, I wasn't ready to admit it just yet.

I fell in love with that beautiful place. I remember my first visit to
Oheka in the fall of 2017, my mom driving as Jeremy and I stared in
wonder at the long, magical, tree-lined driveway. I remember the way
we looked at each other when we first set foot in the library, a space
that captured everything we loved about writing, antique furniture,

and a beautiful time long past. We walked the breathtaking gardens hand in hand and talked about life, love, and how delicious our wedding cake would be. We explored rooms that looked like they were right out of a museum, lit by giant crystal chandeliers, and others that felt like we had transported ourselves to a mansion somewhere, sometime long before either of us got to this world.

Then the wedding day arrived.

Upon reaching the castle, each person was handed an engraved card that read:

Jedediah and Jeremy would like to make their wedding day an unplugged and private event. We kindly ask that all mobile devices be turned off upon entry. Thank you for your participation in making this a truly exclusive evening with our dearest family and friends.

We had done something similar for our engagement party, and so we knew we wanted to create the same atmosphere for our wedding, a time and a place where everyone could have a fully present and engaged experience, without the distraction of technology. We wanted a space where the simplicity of love, conversation, and dancing could flourish. We hired a photographer and videographer to capture the evening but wanted the guests to be captured by the moment.

We exchanged vows we wrote ourselves, before our closest family and friends, in that library we loved so much, amid ivory flowers with warm gold accents. I remember walking down the aisle in my lace ball gown, feeling the warmth and intimacy of the space, surrounded by antiques, artwork, and gorgeous books with tattered bindings that told stories beyond the stories their pages revealed. We danced through the night to music from the 1950s, 1980s, and some country tunes from the 1990s, played by a DJ/band ensemble that included a live singer, trumpet player, saxophonist, and cellist. Everyone was singing along to Michael Jackson, Ritchie Valens, Dion and the Belmonts, Wham!, and more, immersed in a retro vibe that had people smiling, connecting, and just plain having fun. A combination of the setting, the occasion, and the music put us all in a time and place before digital distraction and multi-gadgetry, where everyone was in it together, sharing the experience and the love.

I didn't see one person frantically texting under a banquet table, and when I looked out at the guests to absorb the moment before saying my vows, I didn't see one phone camera, just eyes and smiles and the feeling of engaged, present, in-the-moment, unforgettable peace. The evening finished with a beautiful snowfall in which we ran, played, and danced some more. I won't tell you who won the snowball fight, but I will say that a girl in a wedding dress meant business.

A few days later, Jeremy and I looked back on that night and couldn't stop smiling. It felt like a winter wonderland dream. When we talked about it, what kept coming up was how present everyone was, how the energy of the space felt so connected, so free of the chaos and interruptions we have unfortunately gotten too used to in this tech-inundated world.

Then it hit me. I had done it. I had taken my DeLorean daydream and brought it to life. I had discovered a profound appreciation for the connected, real-life beauty of tech-free moments. And I owed it all to the journey you're about to walk through in this book. I wasn't always so good at removing tech distractions, at appreciating the people and things right before my eyes, at living in the real-life moment. Now I can't imagine my life any other way.

It was a long road to get here. It started with me wanting something I thought I could never have.

MY PARENTS MET IN A BAR. IT WAS ONE OF THOSE SMALL NEIGH-borhood bars on Long Island with real brick walls, low ceilings, and a few big lamps advertising the beers of the 1970s: Schlitz, Schaefer, and such. The place overflowed with young professionals, teachers, and firemen, most of whom flocked there to let go of the outside world for a few hours and absorb the good sounds of local bands that came around on weekends to take their chances on the small corner stage. My mom's roommate sang in one of those bands, so my mom was there on that particular night, standing near the bar, quietly singing along.

My dad says it was about 9:00 p.m. when he walked into that bar. He was on his way home from watching a basketball game at a friend's house, heard the jazz tunes seeping out of the front door, and decided to step in for a drink. Once inside, he glanced around. A slim woman with reddish-brown hair caught his eye right away. He sat down at a nearby table for a few minutes and attempted to attract her attention, but had no luck. She was lost in the music. So, in between songs, he went over and introduced himself.

"Hi, I'm Tony."

"Linda."

"Can I buy you a drink?"

She considered it for a second, then raised her arm so he could see the full glass in her hand. "No, I'm okay, I have a drink," she said, then smiled and turned her attention back to the band, playing a little hard to get.

The band played its next song and he turned his attention to the music, though now and then my dad would look over at my mom and smile. She'd smile back. When the band went on a break, he asked her to join him at one of the small corner tables. She agreed. As the clock ticked along without either of them noticing, they talked about where they grew up, what their families were like, and what they wanted for their futures. Nothing was off-limits. Then, he asked her out on a date. She said yes. So they went out. It was nice. More than nice. Then they went on another, then another, and another, until they were engaged six months later.

"That seems so quick," I said to my mom when I heard the story again several years ago, having had my own dating experiences by then.

MY MOTHER: "Not really. We spent a lot of time together."
ME: "What did you do?"
MY MOTHER: "We talked. We looked into each other's eyes. Most nights it was just us two, figuring out life together."
ME: "No interruptions? Just you two?"
MY MOTHER: "Who would interrupt us?"

She couldn't even imagine what I could be talking about, the concept of a nearby *bzzzzz* with a slew of beckoning alerts not even on her radar. What took my parents months to learn about each other, reading each other's body language, feeling the energy in the space between them, uninterrupted by anyone or anything, would take many people years now. They'd be too distracted juggling an endless onslaught of emails, texts, app alerts, and social media notifications. My parents spent focused, engaged time together, and learned quickly that they were right for each other.

And so their families met almost immediately.

OF COURSE, BACK THEN OR TODAY, MEETING THE FAMILY of the person you're dating can be quite an adventure. My mom grew up in a small downtown Brooklyn apartment with her parents and two siblings, though her brother and sister were quite a bit older than she was and moved out when she was pretty young. My mother's grandma, several aunts, and a cousin lived in a brownstone nearby that was the hub of all family gatherings. My grandmother was one of thirteen, all of whom were big in personality but small in stature—not one of them a smidgen over five foot one. Each one of them was crazier than the last. I'm not kidding. They were completely off the wall, a family of lovable comedians, except for my nanny, the oldest, who took everything way too seriously. Aunt Ana, Milly, Nora, Cousin Didi—the whole crew could make you laugh so hard, your abs would hurt for hours. Sadly, some of the aunts passed away before I was old enough to get to know them, but I'm told they were some of the best comedians of the bunch.

Mom told me that everyone hung out in my great-grandma's kitchen eating, laughing, and talking for hours. It was a small kitchen, simply decorated with a flowered tablecloth and matching wallpaper, copper cookware hanging on the wall near the stove, wooden shelves lined with mementos collected from neighborhood shops. Occasionally someone chased someone around that kitchen with a loaf of Italian bread, which I also saw play out in later years when Nanny would get mad at Poppy and jokingly reach for her long and crispy weapon of choice from the pantry.

Mom's family didn't have much money, but they knew how to make each other laugh and smile a whole lot. The Brooklyn accents were heavy, the Italian food delicious, and the jokes quite colorful. They've told me that the fun banter often carried straight through the night.

On the very opposite side of the spectrum was my dad's family, Italian from Sicily (my mom's side was from Naples), with a more formal, reserved, quieter, American-Italian kind of feel. I don't think I ever heard anyone on that side speak Italian, whereas Nanny and Poppy on my mom's side spoke it quite a bit, including all of the bad words I laughed about later in life when I realized what they had been saying.

When my mother told my grandmother and her aunts that she was going to bring her new beau to meet the crew, she announced: "I'm bringing Tony over tonight. I like this one, and he comes from a good family, and they're kind of proper and all that, so when he comes in, you'd better all behave. No cursing!"

Everyone at my great-grandma's brownstone was warned to be on their best behavior. Mom's aunt Rosie was dressed in her Sunday best, poised to take his coat. Aunt Nora held up a tray of antipasto. Everyone and everything was in its place. But as my six-foot-one-inch dad ducked through the threshold from the entryway into the living room, the first line of greeting he heard was from Aunt Mary, who couldn't help herself: "Wow, Linda! He's so fucking tall!"

My dad laughed, my mom put her head in her hands, and they were all soon eating eggplant parmigiana and telling old, funny stories. Mom's crazy, warm, comedic family made him love her even more.

MY GRANDPARENTS MET IN A SIMILARLY ROMANTIC FASH-
ion, in a way that wouldn't be so easy these days—unless you can bump into true love while shopping online. My poppy, Silvio, and my nanny, Louise, both also grew up in downtown Brooklyn and didn't have much money. As a seventeen-year-old, my grandfather used to run errands for people. He would get paid to pick up a loaf of bread at the bakery, get a box of nails at the hardware store, or bring shirts to the laundry. He'd do those deliveries all over the neighborhood, taking in coins for this drugstore errand down the block or that soda shop run a few streets over.

My grandmother, who was a few years older than my grand-father, did most of the shopping for her family. She went to this grocery store and that market, picking up one thing or another, also all over the neighborhood. As my grandmother and grandfather crisscrossed the streets of downtown Brooklyn, they often stumbled upon each other. One afternoon my grandmother, who was a bit of a boss, turned to my grandfather in the local pork store, right there beside the *soppressata*, and said, "You're going to ask me out on a date."

"I am?" my grandfather answered, surprised but excited.

"Yes."

"When?"

"Now."

"We're going on a date now?" he asked, wide-eyed.

"No, *ragazzo sciocco*, silly boy. You're going to ask me out on a date now for next Friday."

"Oh."

Their first date was at a Brooklyn street fair, with popcorn, cotton candy, games, and all that good stuff. They spent time together, they talked, they walked—just them. They soon were married. They lived, mostly, happily ever after. Italian-bread chases around the kitchen included.

It's no wonder I have a genetic predisposition toward the real and the romantic. When I was younger, I heard these family courtship stories often. The seed was planted and I believed, with all my heart, that these dreamlike encounters would happen for me. I grew up, dated a few good men, a few not-so-good men. I lived. I learned. But soon the gadgets caught hold of my peers and me, and we grabbed hold of them. Suddenly, everyone had their eyes buried in their phones, and I wondered if I would ever find the dreamy I-see-that-lovely-look-in-your-eye interaction that my parents and grandparents had. I started to have my doubts.

I couldn't imagine all of that intimacy surviving now, at least not with another person under forty, as so many twenty- and thirty-somethings were endlessly buried in the tech in their lives. If I wanted old-fashioned romance, I started to think that I'd have to go . . . old. I figured my best bet would be a nice septuagenarian. It turns out, the universe had other plans for me, but before that, my dating life skidded into a worst-case tech scenario . . .

#

WHEN I FIRST TRANSITIONED FROM TEACHER TO WRITER and television commentator, and started working on air, I had several talented colleagues: producers, crew members, writers, and others with whom I felt privileged to work. One was a guy whom, for the sake of confidentiality, as well as keeping with my practice of changing the names of friends and colleagues in this book, I will call Kyle.

Kyle and I saw each other at work pretty much every week. He was friendly and joked around a lot with everyone. In a very tough and stressful business, laughter becomes priceless. Although Kyle was a more extroverted, life-of-the-party kind of guy than the shy Peter Parker type to whom I was typically drawn, he caught my attention.

I have always been a private person. I don't give out my email address or phone number easily. One day, Kyle said he was inviting a few people over to his apartment and asked if he could text me the details. I heard a family friend's voice in my head from that morning: "It wouldn't kill you to be social every now and then." So I gave him my number. He texted me the information. The night of the get-together rolled around and at the last minute Kyle texted to say that many people had canceled, but a few friends were still coming over to watch a movie at his place. He asked if I wanted to join them. I politely declined.

I'll never forget the afternoon I finally said yes to hanging out with him. It was a terrible day. I had gotten my credit card bill in the mail and realized that I was in debt from funding the headshots, wardrobe, and website it would take to launch this new career of mine. I had also had one of the only fights I've ever had with one of my best friends since childhood, about something ridiculous. I was sitting outside on an Upper East Side stoop feeling awful. Where was my career going? Had I been right to give up the stability of teaching? Where did I even want this career to go? How was I going to afford it all?

I was lost.

And just like that . . . *bzzzzz.*

Kyle with a joke. He was that naturally funny guy who could make you laugh without even trying. I don't remember the joke, but it was a good one, and I chuckled. He followed it with some version of "Come on, one drink? One?"

I folded my credit card bill in half and stuck it in my pocket. I needed some fun and funny, and so I was off.

It was an entertaining evening. Completely silly. I met Kyle at a small wine bar and we hung out for a few hours. I laughed. A lot. He seemed to have an endless supply of gossip, too, almost all of it hysterical. I'm not a drinker, but I had some wine that night. I had a glass, maybe two. He had two bottles. I figured he was busting loose from work stress as well. I later learned that, for him, two bottles of wine almost constituted a night on the wagon.

After hearing I'd gone out with him, one of my friends, who knew him better than I did, said, "You must have been really hard up for a laugh." Kyle wasn't the most stable guy and everyone knew it. My friend had spoken the truth. And yes, I was painfully in need of a laugh. But it wasn't just that. Daunted by my current situation and unclear on where I was headed in this new industry, I found Kyle's goofiness and lightheartedness surprisingly refreshing. I decided I needed to spend time with this man. I knew he wasn't the reliable guy, or the quiet, vulnerable type I usually liked, but I didn't care. I said to myself, *I'm just gonna laugh*, and this was the right man for the job.

We started hanging out with each other after work. We went out on five or six dates. I was having a great time. Who doesn't want to sit and laugh for a few hours?

Though things were also weird. For instance, he'd text me late at night, actively and relentlessly, for hours. I wasn't a fan of texting at length, being locked for a long time into words without inflection, exchanges without tone, emojis replacing the sound of laughter or the visual of tears. But I was trying to change my ways, to keep up with the times, to forget my instincts about tech so I could get that laughter I so desperately wanted. So there I was, engaged in the texting, the banter, all the while wondering why he didn't just pick up the phone and call me. I couldn't understand how he could be glued to that little keyboard for such extended periods of time. It felt a little cowardly, too, because the stuff he was saying in the texts often didn't seem like anything he'd venture to say in person. Yet, the laughter was intoxicating, and the next thing I knew, we were in an exclusive relationship.

Soon enough, though, something didn't feel right. I didn't feel like myself with him, surrounded by so many near strangers all the time, constantly in the bar or club scene. About three months in, I realized the extent of his heavy drinking. He said he was from a small town where drinking was just a thing you did when you had nothing else to do, that it totally wasn't a problem. I wanted to believe him, so I did. A month later, I caught him doing hard drugs. Having never done drugs myself or spent any time with anyone who did, I was embarrassingly naïve about it all and soon realized I had likely missed earlier signs. He said it was something he had done years ago and that this was one misstep. I wanted to believe that, so I stuck around. *Stupid girl*, you're thinking. I know, I know. I would be thinking it, too.

Five months in, I started to feel like there was a deeper deception going on. I couldn't confirm what it was, so I let the feeling linger and kept my eyes open. I addressed the drugs and alcohol in depth. He promised to take steps to make it all better. He was still the funniest guy I knew. His family was wonderful and wholesome. When he told me that he loved me or how happy he was that I walked into his life, his smile was warm and seemed so earnest. I kept telling myself that my intuition was wrong. After all, there was no proof of deception, just this dark feeling I had inside. So, I ignored a lot.

Now, I'm not wired for paranoia. Anxiety, yes. Perfectionist tendencies, 100 percent. But paranoia, not so much. I had never worried about a boyfriend cheating on me or even really thought about it. My first boyfriend spent ten months teaching overseas. We visited each other only a few times that year because of my school schedule, but the possibility of him being disloyal never crossed my mind. Growing up, I was surrounded by trusting, committed relationships. I didn't know much else.

What happened next turned that inside out, and me upside down.

KYLE HAD A BLACKBERRY. HE WAS OFTEN MAKING PLANS for us at this bar or that club or this house party. I appreciated his initiative and resourceful energy, and I got used to him being on the phone a lot, making plans. That was us: hand in hand, his hand in mine—until the little red BlackBerry light went off.

One day, one of our mutual acquaintances asked me if I ever checked his phone.

Checked his phone? "No."

I thought that was odd.

She decided she'd overstepped and didn't ask again.

I forgot about it.

Sort of . . .

One night, Kyle passed out at my place after drinking. I was up watching an old movie, hoping it would lull me to sleep. His phone was on vibrate and kept buzzing. Over and over and over. I hated myself for even considering looking at it. But then, in my head, the acquaintance: "Have you ever checked his phone?"

The little red light blinked. My gut whirled.

I picked up the phone.

There were hundreds of texts, countless chains of messages.

Kyle snored.

I opened one of the chains. His best friend was asking him if Kate was still around. Who in the heck was Kate? Someone asked him if he could get some stuff. What stuff? He set a plan to meet in the usual time and place. Where? Who was this? What was going on?

Then another text from someone I didn't recognize, telling him he had left "the goods" in his mailbox. I read on and on. People I had never heard of, talking about a gram, a half gram, a bump, a line . . . Kyle was getting drugs delivered like crazy.

I read the next run, and the next. He was on drugs. He was selling drugs.

Then some earlier incidents that hadn't made any sense started to fall into place. During the workweek, I'd usually go back to my place at night because Kyle smoked a lot of cigarettes and the lingering odor in the apartment made it impossible for me to sleep. He'd text me good night when I'd get home, telling me he was going to bed. As I began to piece together several texts, I realized that Kyle hadn't gone to bed when he told me he had. Instead, he was hosting all-night houseguests, names I did and didn't recognize, girls and guys, in and out, doing drugs, supplying drugs, getting drunk, staying over, you name it.

There were also several flirtatious texts with women I had never heard of, heading to his apartment in the middle of the night, drunk. I had once found a bracelet in his apartment. He had told me it was his best friend's girlfriend's, and that she had left it there when they had visited last. The night I read his phone, I found a text from a girl he had clearly hooked up with, asking about her bracelet, among other graphic texts about the night they had spent together.

During our relationship, there was a woman who would consistently write passive-aggressive things about me on social media. A colleague brought it to my attention, so I discussed it with Kyle. He said he had no idea who she was but that I should have someone keep an eye on it for safety reasons. The night I read his phone, I discovered that he not only knew her but she had been to his apartment multiple times, drinking and doing drugs, and had visited him at work—which, you may recall, was also *my* place of work—and had slept with him more than once.

My heart racing, my hands sweating, I kept reading the texts. There were so many. I couldn't stop. So many plots to keep information from me, so many interactions about getting drugs to or from him, so many sexual exchanges with strange women. The hours went by. Kyle, or whoever he was, snored loudly. I put his phone down. Then he rolled over, deeper into his drunken stupor, and, I'm embarrassed to say, I cried like a baby.

I'm not sure I had ever felt dumber in my life.

My eyes blurry, I took a deep breath. This was madness. I had known Kyle for a couple of years and dated him for months. I knew his family. I had been to his hometown and cooked breakfast in the house he grew up in. I knew he'd messed up a little when he was young, but nothing to signal . . . whatever this was.

Worst of all—and it's still hard to admit this—Kyle had mattered to me. I couldn't, and still sometimes can't, imagine him capable of some of the things I read. Of course he denied it all, even the very words he had written that were sitting right there on a phone that rested on the table between us, words he had typed in black-and-white. He was defiant. Full of excuses. I never got an apology. But once I saw what was going on, I stopped expecting one.

Which was okay. Because this was not a man I knew. His phone was not a phone anymore. It was some disturbing portal to a whole separate existence.

I broke up with him.

It was over.

Except that it wasn't.

For months afterward, when I would see a little red light flicker on someone's BlackBerry, I would feel sick and have to look away. I can't blame the device, in the same way that I don't blame weapons for crimes. I blame people. Things have the power you give them.

Maybe Kyle was going to do those terrible things no matter what. But I also believe that technology aided his downward spiral. It made his lies easier to tell, his deceit easier to hide, his dark desires more easily fulfilled through unlimited outlets.

It's not technology that decides if we are heroes or villains; it's us. But there's no doubt that Kyle's worst instincts were being activated, enhanced, and facilitated by his little device.

#

IN THESE DAYS OF SUPERFAST ACCESS TO EVERYONE ANYwhere, the boundaries around bad behavior are low. That's another reason I miss the good old days, before the sharp, dangerous turn of this century. Back then, if you were going to do something bad, it took time, energy, and effort. When I was in high school, I heard whispers from friends about someone's father, who they said was a cheater. Before the Tech Takeover, it took some work to cheat. It was difficult, a challenge. This man had to leave his house, find a landline or pay phone where no one would overhear him, call the woman with whom he was cheating, arrange a meeting, then find an excuse to leave the house or work and drive far away to some other location, some other town, where he and his mistress wouldn't have to worry about being seen by anyone who knew them. He had to scope out the location, the bar, the hotel, and make sure no one he knew was there, constantly looking over his shoulder to make sure he was in the clear.

In other words, the whole thing took a lot of planning, effort, and even some vigilance. This man had to make choices about what he was doing every step of the way, and in the process of putting in that time-consuming effort, he'd invariably come up against some Is-this-really-worth-it? moments. The risks, the potential pitfalls, the energy—some of these things would surely occur to him as he decided whether to make those potentially problematic moves.

Now? In our current climate of easy access to everyone all the time, the world of naughty flirtation and virtual escape that gives you a little boost of feeling wanted, special, important, needed, is just a Twitter-DM, LinkedIn-connection, Facebook-message tap away. Maybe some once-seemingly-harmless text flirtation grows in that private space, multiplying through the virtual ether, until one day, there you are, in a full-blown emotional affair. You never had to leave your bedroom, your office, your seat on the commuter train or bus. Tap, click, send. The vehicle for these engagements is sitting right in your hand. Then, when the conversation is over, the cheater can quickly and easily delete the text, the DM, the app, even the FaceTime session.

Anyone can swipe here, text "meet me there," and have a clandestine meeting within hours, sometimes minutes. A quick search of the Internet for websites that encourage people to use their technology to secretly engage in another world outside their primary relationship reveals dozens of sites like Fling.com, Naughty hookUP, SnapSext, BeNaughty, and Flirt.com, each of which promises quick, discreet connections. One of the most famous of these sites, Ashley-Madison.com, which was famously hacked in 2015, revealing the names and proclivities of millions of its users, proudly touts the tagline: "Life is short. Have an affair."

The Internet allures, beckons, invites, stimulates. Some seek it out. Some slowly trickle in. Soon enough, too many are embedded. Once again, I'm not blaming the medium. People have the power to control their actions regardless of the situations in which they find themselves. But I'm also not going to ignore that these tech spaces are often designed to facilitate the bad behaviors that were once much harder to actualize. Many of us have lost all sense of appropriateness in our online communications. We say things in those spaces that we'd never say in person. Now, some may argue that virtual flirtation means there's no actual, physical contact and so no cheating. But not everyone agrees on this point. I sure don't.

In May of 2017, Ian Kerner, a couples' therapist, wrote a piece on CNN.com titled "What Counts as 'Cheating' in the Digital Age?" He mentioned a couple that had come to see him after an incident of what the wife called infidelity. Kerner revealed, "He was searching for other women online, she said. Yet [her husband] claimed that he hadn't done anything wrong and that he would never cheat on his wife."

Emotional infidelity has been a slippery slope of morality for generations. When you share a deep, personal confidence with another person of your preferred sex—a confidence you would never share with your partner or spouse—is that cheating? When you enjoy engaging in an activity with that person—an activity that makes you happier than time with your spouse or that you'd rather do with that person than with your spouse—is that cheating? When you have long conversations with that same person—and talk about things that are more intimate than any conversation you have with your spouse—is that cheating? Many people would answer "No," "No," and "No" to all of these examples. Some may say "Yes" to some and "No" to others. The reality is that it doesn't matter what each of us says about what others do. What matters is how each of us, when we are in a relationship, defines the optimal territory and proper boundaries of that relationship. If both people are on the same page, it's a win as far as I'm concerned. The same goes for our online presence and interactions.

My horrifying tech experience with Kyle gave me plenty of insight into what cell phones can hide and conceal in the wrong hands. He had even put people in his phone under aliases, and I can only imagine what he would've done had he been plugged into apps. Maybe he was, for all I know.

In this try-anything-and-everything, YOLO (You Only Live Once) culture, technology is expanding the territory we used to call the Dark Side by making it super easy to explore. Now anyone can wade into the low tide of online pornography, gambling, or drug delivery; engage in virtual worlds with real-life people in all sorts of ways; and even poke their nose into their ex-boyfriend's life with a quick search through Facebook and an even quicker private message. The keyboard courage for many is strong. But thinking, considering, taking responsibility for the potential consequences of our *click-tap-send* urges? Too often, these days, not so strong.

AFTER MY EXPERIENCE WITH KYLE, I LOST SOME FAITH that I'd ever find old-school romance, let alone one-on-one engagement, in the online world. Then I realized that it wasn't just romance that was losing out to these ubiquitous devices.

It was friendships, too.

It happened one night when I was out with Marie, one of my closest friends. We've known each other since middle school. We talk a lot on the phone and get together a few times a year to grab some food, laugh the night away, and forget our stressors, or make fun of them until we are exhausted and relaxed. She's pretty much a sister from another mother. On that particular night, we tried out a new restaurant. We were there for hours, as we usually are, just us, no one else. And yet, we didn't really spend any time together.

What happened? We spent the whole night taking photos of each other, or of the two of us together, or of things around us, and posting them up on social media. We'd taken pictures together before, of course, and even posted some, but on that night it was constant. It wasn't just the pictures. I don't think we had a single uninterrupted conversation. Our phones were out all night, front and center, and alerts were rolling in constantly, the beeps and lights of the gadgets the central activity at our table. Social media notifications—and some crazy texts—took up the whole conversation. I went home troubled, but I couldn't yet process how things had changed so quickly and what to do about it.

I saw Marie again soon after at a birthday lunch with other friends, and it wasn't any better. In fact, it was much worse for me, because I now had my eyes open to it all and had made the decision to put my phone away. I was like the sober friend in a room full of drunk people, the only one seeing things clearly. Phones out everywhere, fingers obsessively texting, cameras clicking to post a photo on social media, a bizarre fascination with how many likes or comments this pic or that one would get.

It had happened. One of my best friends and I had somehow let our friendship devolve to always include a third party.

Her name was iPhone.

The madness was blinding.

I couldn't stop asking myself how this had gotten so bad so quickly.

#

ACCORDING TO A 2017 PEW RESEARCH CENTER FACT SHEET, over 95 percent of the U.S. population owns cell phones and 80 percent of these are smartphones. That's about as prevalent as any technology has ever been in this country. Think about it. Not only does almost every single person in a household have one, but many people have more than one. They're everywhere. They're practically appendages at this point.

Take a ride on the subway, a train, or a bus. Go to dinner with friends. Grab a drink at a nearby bar. Casually stroll down any populated street. You'll see smartphones all over the place. And it's not just smartphones. There are all sorts of personal digital assistants on wrists, on heads, in ears, on glasses.

At one point, I considered getting a flip phone again, removing apps and texting, and that would be that for me. But I liked my weather app, my cycling app, even some social media functions, and I realized that replacing the phone would be a copout, an admission that I didn't have enough self-control to set my own boundaries, a violation of the personal responsibility values I hold so dearly. Instead, I would take control of my life and modify the way I used the device in my hand. We are responsible for what we choose to engage in and how we behave. If I do a bad thing on my phone, it's not my phone's fault; it's mine. I have—we all have—the power to reject the things that don't suit our values or that threaten our peace and happiness.

# #

I DISCOVERED THE ANSWER TO MY QUESTION:
Can I find authentic interaction in an online world?

It was this: *Yes, but first you must take the high road and go into airplane mode.*

I began to prioritize personal interactions by FOCUSING ON REAL ENGAGEMENT.

As one of my first steps, I called Marie. I was honest with her. I told her that I felt like technology was getting in the way of our friendship, that being plugged into all the selfies and texting and social media when we were together left me missing the quality time we used to share. She totally agreed with what I was saying. I was relieved.

I saw her soon after, no phones in sight. We had an amazing day visiting a cat café where adorable cats of all ages roam around and you can pet and play with them. We had lunch on the Brooklyn Heights Promenade and explored some open houses for brownstone rentals that left us in complete awe of old architecture. We got our friendship back because we both stood up and fought for it.

As a next step, I shut off the beeping, buzzing notifications and put my phone away when I was with other people. I took it off the conference room table and the dinner table. Sometimes I never had it in the room at all. If other people I was with socially were glued to their devices, I'd suggest we meet later or another day when they had time to talk.

Of course, this took some time. My journey to understanding how to take the high road and go into airplane mode was a long one. I didn't fully recognize how the tech tumult was affecting my life until I encountered that last straw, the fiasco with Kyle. Up until then, I'd gone through a series of mishaps, including manic panics, distress, and the messed-up feeling of compare and despair as I spent too much time staring at social media and not enough time contemplating sunsets, all of which were a result of my tech-polluted life. Worse, I didn't even know what was happening to me. Some of the moments were big, some were small, many were, I later discovered, universal experiences. I was not alone. Neither are you. That's why I reveal them in these next several chapters.

Fortunately, my romantic journey had a happy ending in a castle on Long Island. In fact, it was when I'd hit rock bottom in my tumultuous quest for a resolution to all this tech overflow that I met my husband. And he wasn't seventy years old. In fact, he was younger than me. Heck, he was downright millennial. But how Jeremy and I met and fell in love is a whole other wonderful story I'll get to later.

Bad people are going to do bad things no matter what. Kyle was going to deceive me anyway, I'm sure. The phone just made it easier. Good people can be good people with smartphones if they take the time to think about their usage. Then, perhaps, we would all have healthier relationships with our phones, the people in our lives, and ultimately ourselves.

First, though, I had to admit that not only was technology overwhelming my friendships and my love life, but it was also eating into almost every aspect of my day.

CHAPTER 2

CREEPING, CRAWLING, APPALLING ADDICTION

What if the things that seem to make life
better are actually making it worse?

I WAS HYPERVENTILATING. NO JOKE. I COULD NOT BREATHE.
My palms were sweaty, my heart was racing, the ground seemed to
swirl beneath my feet. You would've thought I had suffered a trau-
matic event, like losing my Maltese or some other living, breathing
being I'd adopted, or who shared my DNA.

But no.

An old college friend, whom I'll call Jane, had been in the backseat of a New York City taxicab with me. It was a Friday. We were
looking forward to a night out with our crew of buddies who had
been trying to organize a get-together for months. We were finally
going to meet up. This was the night.

The yellow cab meandered through Manhattan, the impatient
driver briskly navigating the narrow streets of Greenwich Village,
then turning up Tenth Avenue. Buildings and people blurred past
the cab's windows, and Jane and I were relaxed and comfortable in

the back seat, each glad to be in the trustworthy company of an old friend, decompressing after a week of work stress.

It was the spring of 2012, and a presidential campaign season was under way. I was doing a lot of television appearances on Fox News, and Jane and I were caught up in a conversation about how in the heck I wasn't losing my mind talking politics all day long. Jane knew that I don't like to talk politics socially and that, in fact, when I'm with friends or family, I'm usually the one who raises a firm hand and declares a "no politics" zone. As the taxi continued past the Meatpacking District, through Chelsea, and up into Hell's Kitchen, we were ranting about the frenetic media storm that was the 2012 news cycle and how exhausting the polarizing prattle of politicians could be. We hardly noticed that we had crossed Central Park and arrived at my apartment on the Upper East Side.

The driver pulled over and clicked off the meter, lighting up the numbers on his roof that signaled his availability. An eager traveler rushed from a bodega to grab the cab for himself. He was wearing an alligator costume, complete with a long tail that dusted the concrete. New York City never fails to disappoint. We paid quickly, stepped out of the cab, and the driver and his new fare zoomed off.

As we entered my building lobby, I indulged in my inclination to check my cell phone for the umpteenth time. I reached for it, and . . . *No . . . this can't be . . . wait . . . oh no* . . . It wasn't . . . anywhere . . . to be found.

Yikes. My stomach dropped. My phone was gone. I'd left it behind. It must have slipped out of my hand, my pocket, my bag. I couldn't recall. All I knew was that . . . I didn't have . . . *gasp* . . . my phone . . .

Breathe, breathe . . .

I was panicking. Big-time.

I whirled around to look for the cab, thinking that maybe the driver was stuck in traffic or at a light, and that I could flip off my heels and chase him down. Nope. Nothing. No sign of him anywhere. I'd left my phone in a cab on the one night in forever that the cars in Manhattan were moving smoothly. He was long gone.

Jane recognized my panic and acted quickly to help. We hadn't gotten a receipt from the driver, so we didn't know the number of the cab. Since we had paid with a credit card, we hoped the transaction and our trip could be tracked.

Go, technology.

Jane called the New York City Taxi and Limousine Commission's lost-and-found hotline. They said they'd look into it and get back to us.

But that meant nothing to me. I was untethered, floundering, at a loss. I couldn't function without that little electronic device that somehow held my whole world inside it.

"What time is it?" I asked, apparently unable to even tell time without my phone.

"Six thirty," Jane said. "We've got to . . ."

Jane went through the litany of what was ahead of us, what we needed to do to meet up with our old friends for the long-anticipated evening. It was going to be so nice. We were all going to go grocery shopping, prepare some delicious food together, talk and laugh a lot, and then enjoy a nice black-and-white film, oohing over Cary Grant and aahing over Loretta Young. Though, for me now, the evening had taken on a dark tone, my thoughts careening off in distracted directions, the potential of the get-together lost, anticipation and excitement drowned out by my overriding unsettledness.

"We'll go to the market and get—"

I grabbed her arm. I couldn't do it. I was too paranoid. Where was my phone? Did someone have it? Was someone reading my texts or emails? Who needed me? What was I missing?

"Seriously?" Jane asked.

"Seriously. I don't think I can do anything until I get my phone back. I don't know who has it or who's trying to reach me or—"

She stopped me there. She understood.

We canceled the night. Our friends didn't meet us. There was no dinner, no movie. Just Jane, me, and my anxiety, pacing my apartment, waiting.

Boo, technology.

Or at least my attachment to it.

Finally, the door buzzer rang. The nicest of drivers had received our message, retrieved my cell phone, and brought it back to me in one piece. More than grateful, I tipped him again, this time over three times the fare, and he left.

The sleek rectangle firmly back in my grip, I was smiling ear to ear. The endorphins zoomed as if I'd been reunited with a missing child I had lost at the mall. Relieved and giddy, I looked at the phone and noticed I had missed six texts. I went right to them.

From a friend: "Hey what's up." From another: "Yo." And three other insipid "Thursday? Lmk," "Saw insta lmao," "WTF did he do to his hair?!" Oh, and another all-important "What up."

Her hand on my shoulder, Jane turned to me, concerned. "What did you miss?"

I looked up, embarrassed. "Nothing. Except three hours of my life."

She nodded and raised an eyebrow. "Yeah. I missed them, too." She still loved me, but I had also ruined her night. And I knew it.

"Could've been a fun time," I said.

"Yeah, could've been," she agreed.

I WASN'T ALWAYS LIKE THIS. IN FACT, I GREW UP DOWN-right normal.

My family was a pretty standard Italian-American, hug-a-lot, talk-a-few-decibels-above-normal, plan-your-next-meal-while-you're-eating-the-current-one family. My parents, who each came from larger families, got together and had one child: me.

Our family was old-school. Even though I grew up in the 1980s and 1990s, my parents still kept a small black-and-white TV in the kitchen, the old-fashioned kind with a channel dial and an antenna you had to move around when the picture on the screen got fuzzy. We had color televisions and cable in the bedrooms, but even those TVs were older models with press-in channels that had been in the family for years. It took some maneuvering to adjust them to cable remotes. We had a phone with a rotary dial and a long curly cord that could stretch around corners as my mother went from one room to another, chatting with an aunt or her sister. It wasn't that my parents couldn't stretch to afford the latest and greatest innovations. It was just that they didn't see the need for them. The only modern appliance we had was a microwave. When you have leftover lasagna staring at you a lot, trust me, you want it to heat up as quickly as possible. Oh, and a VCR, because Mom likes old romantic movies and Dad likes old horror films. I came to like both.

I watched a fair amount of TV growing up. My appreciation for self-reliance, personal responsibility, and the simpler things in life was forged early on, influenced in no small part by my regular viewing of my favorite shows: *Wonder Woman* and *Little House on the Prairie.* I balanced that with outside time, hanging out with kids on the block, playing kickball at a nearby park, and exploring the nooks and crannies of our neighborhood, from playgrounds to parking lots to each other's rooms. I was into school, a few close friends, and my parents. And, of course, all the middle school fashions. I was one of the many preteens guilty of begging, pleading with, and eventually overwhelming my hardworking middle-class parents to acquire the entire line—each color—of suede boots that any respectable Staten Island girl had to have. In my case, that meant almost a dozen. Despite that wrong turn into Imelda Marcos–like conspicuous consumption, we were a simple family, with a few splurges, such as our annual vacation to visit my parents' favorite escape spot near Sanibel

Island, Florida, and a trip abroad together once when I was fifteen. The three of us ventured off to Belgium and France for an exciting, tap-into-the-savings, once-in-a-lifetime adventure. My dad took with him a basic video camera he borrowed from a cousin. No bells, no whistles.

I look back at those days and there was a calm. A stillness. A pause.

I liked that. There was time to think and reflect and imagine. I can't help but feel that our brains needed that time to recharge and regroup.

Where has that time gone?

It's been a while since I've seen a child hang out in a park, crouched over a puddle after a rainstorm, stick in hand, tracing through the surface area, watching the water ripple toward the edges, thinking, daydreaming. Or a kid on an airplane staring out the window, intrigued by the movement of the baggage handlers coordinating the freight on the tarmac. Nowadays, what do I see? Children glued to big tablets in their hands, clueless to their surroundings, entranced by the make-believe, engaged in a process that pulls them mindlessly along a predetermined trail engineered by some Silicon Valley twenty-something.

But alas, who am I to judge those kids? I only started noticing so much of the world when I finally put my own phone down. The me of my olden days would be shocked to meet the me of now, with my cell phone and laptop and flat-screen TV. As Michael Harris notes in his book *The End of Absence: Reclaiming What We've Lost in a World of Constant Connection*, "You live in an ecosystem designed to disrupt you and it will take you for a ride if you let it."

WE HUMANS GO WAY BACK WITH OUR NEED TO IMPROVE our lives with innovation. As V. Gordon Childe notes in his 1936

book, *Man Makes Himself,* "Man's emergence on the earth is indicated to the archaeologist by the tools he made." In other words, the very proof of what we were, and how we got here, comes from our creation of technology.

The first tech consisted of tools for gathering and consuming food and water, making weapons and clothing, and capturing fire. This was followed by equipment to help us farm, corral animals, garden, and build. Then the wheel was a big deal, of course. As were engineering and architecture, pen and paper, locomotion, plumbing, music, roads, the printing press, medicines, textiles, metals, bicycles, railways, electricity, radio, telephones, automobiles, airplanes, vaccines, refrigeration, automation, radiation, television, rockets to the moon, personal computers, the Internet, cell phones. And now . . . smartphones, laptops, tablets, Alexa as some responsive mechanical resident in many homes, robot vacuum cleaners, virtual reality gaming, 3-D amusement park rides . . . I could go on and on, because we live in a world where developments of new technology are introduced almost daily.

I've always liked technology. I mean, who doesn't like electricity, air-conditioning, and that wonderful vacuum package sealing machine that keeps my favorite sea-salt-and-vinegar potato chips crisp? In fact, I'm sort of, kind of, a product of technology, because my very name comes from a 1970s TV show.

Though my mother hates when I reveal that.

My name is Jedediah. Which, though it seems easy to pronounce ("Jed-ah-dye-ah"), isn't for everyone. Some of the folks at Fox News call me by my initials, J.B., and several of my former colleagues on *The View* say Jed. But I also had a nickname when I was little, and that was A.J., because those were my initials, A.J.B.

NORMAL PERSON: "What does the *A* stand for?"
ME: "Nothing."
NORMAL PERSON: "Nothing?"

MY MOTHER: "I wanted her to have the same monogram as her father, and he's Anthony James Bila."

NORMAL PERSON: "So why not give her a first name that starts with *A*?"

MY MOTHER: "I wanted her to be Jedediah, so I just put an *A* in front of it, like F. Scott Fitzgerald."

ME: "Mom, his *F* stands for Francis. My *A* stands for nothing."

MY MOTHER: "Well, the *A* stands for Anthony."

ME: "My first name isn't Anthony."

MY MOTHER: "But in spirit."

[NORMAL PERSON: goes silent, staring blankly.]

Cue my mother smiling at me like she totally gets what she's saying.

Cue me staring back at her like I totally don't.

Frankly, the *A* is a pain in my ass. Especially at the airport. The DMV didn't properly understand the *A*, so they printed my name wrong on my license and turned my first name into "AJedediah." Then my ticket reads "Jedediah" because I often forget about the *A* altogether. It's a hot mess.

Cue the airport security guard staring at me blankly.

If I had a dollar for every time someone asked me, "What does the *A* stand for?" . . .

Some people think the name Jedediah is sophisticated and exotic. Which Mom likes. And I like, too. Others think it's a biblical name. Which it is, albeit spelled a little differently. Only my mother didn't get my name from the Bible. She took it from a character on the television show *Barnaby Jones*. A male one, to boot.

MY MOTHER: "Why do you have to tell people that?"

ME: "Because it's true."

MY MOTHER: "I know it's true, but you don't have to tell everyone."

But of course, I'm so frustrated by my most recent what-does-the-*A*-stand-for? exchange with heaven knows who that I wind up telling the world.

Cue me smiling like I totally get why I tell everyone.

Cue my mother staring back at me like she totally doesn't.

The week before my wedding in 2018, while looking at legal documents that had all somehow spelled my first name differently and created a headache, I decided I would soon file the paperwork to legally drop the *A* altogether. I still haven't told my mother.

# #

AFTER 2004 WHEN FACEBOOK CAME OUT, 2006 WHEN TWIT-ter emerged, 2007 when smartphone technology accelerated, and this subsequent decade in which we saw social media overwhelm our very existence and these magic machines land seemingly fixed in everyone's hands all over the world almost all the time, this fan of technology needed to take a look at what exactly she was fanning over.

And so, a positive and negatives checklist of sorts:

Yes, we have easy access to information and entertainment. That's a positive.

But more people can run an app than run a hundred yards down the street. Bodies are going soft. People are sitting around hunched over, looking at screens, consumed by content. Brains are being out-sourced, with easy-access information via constant clickbait replacing the art of critical thinking. That's a negative.

Yes, we can communicate quickly with each other, share our insights and thoughts with others all over the world, and give platforms to those who have traditionally had no voice. That's a positive.

But that means anyone anywhere can anonymously spew their hate-filled messages at anyone, anywhere, in front of everyone, everywhere. That's a negative.

Yes, we are moving toward faster, more efficient, easier everything. That's a positive.

But sometimes a faster, more efficient, easier everything isn't a positive. And that's a negative.

Without some reflection on all this, we stand to lose ourselves and so much more. Personal responsibility is my motto in politics and in life. Technology, like many other things and people we encounter, only has as much power as we're willing to give it.

That's the point of this book.

To become aware of and think about our current relationship with technology. Who's in the driver's seat?

To be honest with ourselves about the role of tech in our lives and how we feel about its effect on our minds, our relationships, our priorities, and our mental health.

To recognize and change the negative consequences of our gadgets and devices and emphasize and support the positive things they bring to the table. We created these machines, which means we also have the power to make them work in a way that is best for each and every one of us.

I knew it was time for me to recognize that a lot of the tech that claimed to make my life better was actually making it worse. It took only a little bit of research to discover that I wasn't imagining it.

A global study published in 2016 by the National Institutes of Health (NIH) found: "The problematic use of cell phones has been associated with personality variables, such as extraversion, neuroticism, self-esteem, impulsivity, self-identity, and self-image. Similarly, sleep disturbance, anxiety, stress, and, to a lesser extent, depression, which are also associated with Internet abuse, have been associated with problematic cell-phone use."

That can't be good.

But here's where it gets downright unsettling. The journal noted that some experts "consider cell-phone addiction to be one of the greatest addictions of the current century."

Wait, what? How can that be? In an era where the news is filled with alarming stories about our country's opioid usage and prescription drug overdoses, researchers are saying that CELL PHONE ADDICTION IS ONE OF THE GREATEST ADDICTIONS OF THE CURRENT CENTURY?

Yes.

The study concluded that "excessive attention and uncontrolled dedication to one's cell phone is an addiction." Addiction was defined, "in its broadest sense, as the capacity to get 'hooked' on reinforcing behaviors, excessive worry about consumption or behaviors with high positive reinforcement, tolerance, loss of control, and difficulty in avoiding said behavior, despite its negative consequences."

That sure sounds familiar to me.

The report goes on to clarify that "defining elements of behavioral addictions [include] the loss of control, the establishment of a dependent relationship, tolerance, the need for progressively more time and dedication, and severe interference with daily life." It adds that "these behaviors lead to uncontrollable use, in addition to feelings of intense desire or irresistible need, loss of control, inattention to usual activities, the focalization of interests on the behavior or activity of interest, the persistence of the behavior despite its negative effects, and the irritability and malaise associated with abstinence."

It all reminds me of how I felt when I left my phone in the taxi.

Anxiety . . . stress . . . depression . . . loneliness . . . insomnia . . . very real consequences of Tech Overload. With more and more people on their phones, younger and older all over the world, this means many, many, many people are infected—I mean *affected*—by the ever-present usage of tablets and cell phones. A Common Sense Media poll found that "50 percent of teens 'feel addicted' to mobile devices." And a third of the families polled "argue about it daily."

All of this feeds my concern that the prevalence of personal technology products is having the biggest and most enduring effect on the next generation. The human brain operates at high capacity during adolescence. When we are younger, we can absorb a lot of information fast. This means it can be an extremely productive time. But it also means it can be a time when we are susceptible to dangerous stimuli that attract our attention. That's nothing new. See William Shakespeare's *Romeo and Juliet*. A drug by any other name . . .

#

TECH ADDICTIONS ARE A TRICKY BUSINESS. THE DESIRE TO always be plugged in builds and builds until one day you wake up and you're buried in it. Done for. Consumed. Even if my cell phone habit wasn't an all-out addiction, it was something close. An obsession at the very least. A compulsion that was hard to ignore. That's when I knew, there was no doubt about it, I had fallen into a black hole I believe many of us have, what I've decided to call:

OCTD—Obsessive-Compulsive Tech Disorder

OCTD is stealthy. It's the boiling frog metaphor. Put a frog in steaming hot water and he'll hop out. Put him in a pot of regular water in which you slowly turn up the heat, and he won't notice the gradual increase in temperature. Before he knows it, he's boiled, kaput, submerged in the death throes of his inattention. This is happening to us, too. It's a *creeping normality*, which is the idea that something strange and new can come upon us inch by unnoticeable inch until it's the new normal. Cell phones are the hot water, we are the frogs, and we are all gradually, carelessly, mindlessly boiling.

That's unusual for us humans. Our primitive instincts tend to notice when we are in danger. Our hair stands on end, our skin perspires, our Spidey senses tingle. Yet, somehow, as the years have tiptoed along, we've allowed this nonstop immersion in the distracting, alluring, draining world of our devices and come to ignore the warnings of a technology takeover.

Why aren't we asking more questions?:

What is the impact of tech on our lives? What skills are we gaining and losing? Are we happier? Smarter? As Nicholas Carr noted in his book *The Glass Cage: How Our Computers Are Changing Us*, automation of skills like analysis has the upside of "relieving us of repetitive mental exercise," but it also "relieves us of deep learning."

That scares me. Mostly because we are allowing all these computers to replace our plain old thinking brains. We simply are not as alert and proactive when new things show up. Either we're not paying attention, or the creeping normality is taking us so much by surprise that we don't even have the wherewithal to react.

How soon until we outsource one of the very best acts of being a human being, that of being a friend? In her book *Alone Together: Why We Expect More from Technology and Less from Each Other*, MIT professor Sherry Turkle explored examples of those who are working to replace human jobs and actions with robots and technology, including the creation of "lifelike," "sociable" robots. In her observations, Turkle stated that she thinks "social technology will always disappoint because it . . . promises friendship but can only deliver performances." She added, "Do we really want to be in the business of manufacturing friends that will never be friends? . . . A machine taken as a friend demeans what we mean by friendship. Whom we like, who likes us—these things make us who we are."

Professor Turkle is one of many out there researching the effects of technology on humanity. But are we hearing these observations? Are we discussing them in our private lives? Or are we viewing these studies and this research from a detached perspective, like they're some academic exploration whose content and implications don't pertain to our personal, real, intimate spaces and interactions?

It used to be that there would be protestors against the latest and greatest innovation, people asking tough questions to challenge the roles that the newest machines would play in our lives: Would access to so many TV channels make us spoiled brats? Would video games turn our kids into glazed-over, inactive couch potatoes? Is it safe to blast music right into our ears?

Even the ancient Greek philosopher Socrates argued in *The Phaedrus* that the invention of letters and writing would "create forgetfulness in the learners' souls, because they will not use their memories; they will trust to the external written characters and not remember of themselves." In fact, he was so committed to this idea that the only reason we know he said that, or anything else, was because, ironically, his favorite student Plato wrote it down. Can you imagine what Socrates would have thought of the notes or recording apps in our phones, or the practice of outsourcing our knowledge to the Internet?

Socrates questioned the Next Big Thing. By doing so, he made us think about navigating a world with new inventions. People used to challenge the latest gadget, not necessarily to destroy it, but to be aware of its potential dangers and to put it in its proper place in the grand scheme of our lives, our families, our universe.

Where are the tough questions?

Where is the challenge to the machines?

Where is the rebellion?

My close friends aren't surprised that I'm speaking up against the dominance of the machines. And not just because I've been romanticizing handwritten love letters and horse carriage rides since middle school. You see, I was also fascinated by the *Terminator* movies in my youth (and still am), mesmerized by the idea that humans, with all of our limitations and emotions and fragility, could somehow challenge manifestations of technology that seemed so much bigger and stronger than us. I might've dressed up as Sarah Connor for Halloween. Twice. Okay, maybe three times. Although some of that adoration had to do with her ability to do pull-ups in a way I envied, more of it was because she represented the power of the human spirit and the drive to overcome. I realized at a very young age, while watching the first two *Terminator* movies with my dad alongside a mountain of snacks, that humans are fully capable of creating something that will seek to destroy us, and that only we can stop it.

Today, the concept of an artificial intelligence network taking power over the planet isn't so far-fetched. In the summer of 2017, a Wisconsin company called Three Square Market announced that they were "offering implanted chip technology to all employees." Apparently, that company has the ability to implant people with radio-frequency identification technology to buy things, open doors, turn on office equipment, log in to computers, trade business cards, and even store health care information. It makes me want to fly to Wisconsin, stop everyone on the street, and shout: "Haven't you seen the *Terminator* movies? Are you not getting where this could go? Where it's going right now?"

What happens if we give these machines too much power? Already, our computers are often smarter than we are. Our favorite music is systematically selected into a playlist tailored just for us, airplanes can fly themselves, robot vacuums clean our houses. It's so strange that people can grow up watching movies and reading books that flash warning signs of what could happen, and yet don't notice the real-life red flags when they emerge. Why aren't we seeing this? Why aren't we shouting from the rooftops? Or why aren't we at least just a little worried and asking some important questions?

I'll tell you why: because it's becoming the new normal in a subtle, seductive, intoxicating, quick transition. Each time we unconsciously buy into the new normal, it builds up speed, faster and faster, until each new day is nothing like the last.

The frog should start leaping. NOW.

#

I DISCOVERED THE ANSWER TO MY QUESTION:
What if the things that seem to make life better are actually making it worse?

It was this: *When things that were supposed to help you start to hurt you, reconsider their role in your life.* For me, that meant it was time to admit I had a problem and become aware of my addiction.

I realized that AWARENESS IS POWER. Why aren't we more aware of the negative consequences of so much of this pervasiveness of technology on our lives? Are we too tired? Or numb? Have we given up? Are we blinded by some new normal that's not so normal at all? Or does it not matter to us? As I've said, I'm a fan of many things about technology. For a while, I tried to convince myself that maybe there was nothing to rebel against, that maybe it was all okay. Only, that didn't seem quite right because something felt wrong. Especially with the way we are all so accepting of the sheer volume of new technology and the intensity with which it is sweeping over us, like a soothing yet smothering baby blanket.

Humankind has always been inventive, and often the results of our mental maneuverings help us to become more productive and capable. However, historically, introductions by entrepreneurs were met with, at the very least, some kind of resistance, an awareness that things were indeed changing, a questioning. Take the nineteenth-century rebellion of the Luddites. The Luddites were a group of weavers who wanted no part of the new automated, mechanical looms. They fought tooth and nail to keep those new machines out of their cottage industries, but the machines won and the only thing that endured from their efforts was the association of the Luddite name with someone who resists technology.

This fear of human workers being diminished or even dismissed altogether is still prevalent in factories all around the world, yet it mostly seems to be a quiet kind of fear. Where is the significant verbal resistance? The visible turmoil? The rebellion against a system that encouraged it? Amid our quiet fear, the machines—and those creating and enabling them—are just doing their thing, until, with creeping normality, the technology takes over.

We've seen it in many places. Where once there were rows and columns of card catalogs in which to dig through the Dewey Decimal System, now there is one computer on the information counter that can locate any book in the world, let alone in that library. Travel agents are scarce, overtaken by websites. Automated cashiers and tablets in place of waiters are creeping into the restaurant tech space. You can even go through a car wash without talking to a single human being. Just tap the module on one end and drive right on in. Obviously, you and I could cite dozens of examples.

Although there are jobs to be preserved, people's livelihoods to consider, and an active school of thought that believes humans are integral to the thinking and creativity part of the equation, automation of pretty much everything is possible. One hundred–plus years ago, no one could imagine that the car would ever replace the horse and buggy, and now those cars will soon be replaced by automobiles that are truly auto, as in automatic, systematic, self-driven, without even the need for a human at the wheel—heck, without the need for the wheel at all. Change and innovation are happening everywhere with everything, the life cycle of the latest and greatest playing out faster and faster. Nothing and no one is immune. Just as surely as no one ever thought the radio would supplant the phonograph, or the television the radio, or the Internet the television, those things of which we can't even—don't dare to—conceive are sure to come next.

We can't stop the tidal wave that is technology. All we can do is stay aware, speak out, and make sure that we're in the driver's seat the whole time. Too many of us are forgetting who we are and what we value. Too many of us aren't pausing to look at our lives, at what's already been lost, at what could be lost tomorrow or next month or next year by immersion in devices and apps and whatever the next hypnotizing tech development might be. Too many of us are missing out on the best things about being a human being. I see it every day.

I saw it one nice winter afternoon in a café, when the universe sent me a little love lesson.

CHAPTER 3

THE MULTITASKING MYTH

Does anyone make eye contact anymore?
Does it matter?

IT WAS A DAY LIKE ANY OTHER—BUSY, OVERWHELMING, with an endless to-do list—in a season like no other, that gorgeous time between Thanksgiving and Christmas when New York City is full of lights, Christmas tree sales line the sidewalks, department store windows showcase glistening decorations, and old-fashioned holiday music echoes through stores and in taxicabs. It was that time of year when people smile more than usual and you find yourself daydreaming about winter cottages, fireplaces, and all that good, old-fashioned stuff that warms you up inside. The holidays in the city are usually so exciting, my senses wide-awake, my imagination sparked, the shimmering, glimmering, glistening stimuli keeping me aware, bright-eyed, and attentive to the world around me.

After an afternoon of hectic meetings following two live segments on Fox News at the News Corp headquarters, I decided to switch up my routine. Instead of heading to my apartment in Midtown, I would walk to the Upper West Side. My brain was cluttered with news stories, column ideas, and pitches for new projects, and I was feeling scattered. I needed a moment to breathe.

A small café caught my eye and I fancied the chance to relive a time when I used to go to places like that more often, a time when I felt less disjointed, when things seemed simpler. Maybe I felt some nostalgia for being a young adult with fewer responsibilities and the luxury of roaming aimlessly, or for the days of college when I'd pack up my books and head to a quaint coffee shop with nothing but some music, the outdoor landscape, and a journal to write in. Either way, I knew that I needed to take a pause, and I hoped that this place would be the perfect portal.

I sat down at a corner table and ordered what I thought would be the ideal prescription for busy: chamomile tea. However, in no time at all, my phone and I renewed our codependent relationship, my head buried in the usual cycle of texts, emails, Twitter, Instagram, Facebook—UGH.

I caught myself just before my brain spun off the road into its daily ditch of endless distraction. What was I doing? I'd come in here to get away from all that. Yet, there I was, right in it, on it, at it again, plugged in and checked out.

No. This wasn't me. I knew that. I needed to shift back to normalcy.

I put my phone facedown on the table, grabbed hold of the ceramic mug of steaming hot tea, and sat back in my chair, ready to observe.

I noticed a young couple, a man and a woman in their twenties. Something wasn't quite right. I watched for a bit.

I was appalled at how long it took me to recognize what was so disturbing. These two had not, in the five minutes I was watching them, looked at each other. Not once. There was no eye contact. Why not?

Because they were on their phones.

In fact, she was on two phones.

An investigative commentator by trade, and a seasoned, unapologetic eavesdropper by habit, I focused in on their conversation. They were at least talking to each other. He mentioned a friend's Instagram. She read aloud a text. They laughed at a Facebook post, or a Snapchat image, or whatever tech distraction was sitting conveniently between them. Their conversation, which was more *at* than *with* each other, was all about the external content from their gadgets.

At one point, her shawl fell off and became entangled beneath her chair. She tried to retrieve it, her eyes still on the phone. Only after a minute or so did her struggle become apparent to him and so he, eyes still on his phone, reached over and helped her extricate the shawl from the table leg. He made a comment that revealed they were on a date and gently put it back over her shoulders. She thanked him. Yet, even this moment was executed with almost no eye contact. This young couple was on a date that looked nothing like a date.

I glanced around the room, hoping to come upon something different, something hopeful. Another couple's laughter caught my attention. A man and a woman in their late sixties, early seventies. Each held tight to a mug from which they sipped intermittently, intent on the other, only breaking eye contact to make sure their forks made their way to the piece of pastry they were sharing. They looked at each other like they had found something in life they had been waiting for. He told a story. She smiled and took it in. She said something funny. He laughed. The waiter came by the table to check on them, and just as quickly removed himself, the energy between them exclusive, a barrier to the outside world. No one else was invited to that intimate moment but them. Not a phone in sight. No buzzing from a device beckoning their attention. Just two people on a journey whose magic I was lucky enough to witness from a small table nearby.

Two dates.

One where plenty of outside voices were invited, including all of their social media friends.

Another where two people sat immersed in nothing but each other.

ONE OF THE LESSONS I RECALL FROM A COLLEGE ART HIS-tory class is that when you look at a painting, you should be conscious of how it makes you feel. What emotions does it conjure inside you? What part of your life does it make you want to run *to* or run *from*? Art can certainly be a way for artists to discover themselves, but the treasures they create can also help you find *yourself.*

If I were to imagine the older couple in the café as a canvas, I'd feel warmth and intimacy. Tenderness. Vulnerability. It was a safe space. It reminded me of cushy couches, delicious food, and those holiday rom-com movies we find ourselves clinging to in our most fragile moments. Their metaphorical canvas inspired me to want a relationship, to have a life partner, to be in love.

When I looked at the younger couple as if they, too, were a painting, I felt disconnected, worried for their future, worried for my own, and sad that they were missing so much of each other in each passing moment. There was a shallowness that made me think of cold, empty spaces. All I could see were those little buzzing devices that had the biggest, brightest seats in between them. I wondered what they might've discovered had they looked up at each other, into each other's eyes.

I was unnerved. Troubled. And yet, ironically, hypocritically, just minutes before, I had been behaving the same way as the younger couple, my head buried in the tech distractions of my smartphone. I, too, had missed my opportunity to engage in this special time with my chamomile tea, a view of the city, and my own thoughts. Jolted, I tossed my phone in my bag. I was part of the problem. But I wanted to be part of the solution. I was going to fix this for myself.

I looked outside for the first time since sitting down. Out through the window, the business of Manhattan was winding down. The streets were lined with trees, their branches outlined with the transcendent golden rays of dusk. Two tiny dogs passed, one with red booties to protect him against the salt on the sidewalks, the other wearing a baseball jersey. A mother and her toddler held hands, the child smiling, looking up at his mom, and the mother . . . on her phone.

UGH.

It was time to go. I touched base with my producers, gathered my assignments for the next day, and decided, on my way home, that I would unplug for the rest of the night. That's when I saw:

Almost everyone on the subway with their heads down, eyes glued to their phones.

On the street, a couple holding hands . . . and then letting go to look at their phones.

A playground where most of the kids sat on the concrete, each set of eyes glued to a device, no conversation or typical playground activity in sight.

There were phones everywhere.

I started to understand that we humans, walking through the complexities of our present lives, had made a costly compromise. We were losing something beautiful, the simplicity and realness of real-life moments and the discoveries that come with them, in exchange for an "easier" life consumed with tech achievements of accessibility, efficiency, and a supposed broadening of our minds and interconnectedness of our worlds. Quite often, we don't even have a seat at our own table.

Later, as I dug into the how and why of what was happening, yet still couldn't fix my own actions, I became aware of the duality of intention and action, of claiming to want one thing but actually fostering the opposite. I learned that this was an issue for many. As William Powers, in his book, *Hamlet's BlackBerry: Building a Good Life in the Digital Age*, so accurately observed, "For years now, news outlets have been dutifully covering what's been aptly called the Too-Much-Information Age . . ." Powers added, "At the same time, the news media push nonstop connectedness as eagerly as any Silicon Valley titan, and for the same reason: business demands it." A media that warns against something . . . while promoting it. Hmmm, many of us aren't that different. Much like the media, we have talked the talk, warning of technology's troubles . . . while holding an iPhone, laptop nearby, social media notifications snatching our gaze.

I wanted that to change. But being the freedom-lover that I am, I had no desire to tell anyone else what to do. There was also no way I was going to advocate for signs, proclamations, rules, or legislation to fix the problem of a ballooning tech industry that was taking over so much of our lives. I continued to wonder how I could best set the example of being personally responsible for finding a responsible way to manage the ever-growing tech that was consuming our days and nights.

While I was thinking about how drastically our world had changed in the past three decades, I came across an article in the *New York Times* about a current production of a 1990 play, Scott McPherson's *Marvin's Room*. The creative team was discussing why they had to make sure it was clear that the play was taking place in the past, not the present:

DIRECTOR ANNE KAUFFMAN: "We're setting it in the '90s . . ."
ACTOR JANEANE GAROFALO: "It would change everything if there was social media. The kids would be on their phones the whole time."
KAUFFMAN: "There have to be real reasons for them to not actually be in touch with each other for 20 years."

ACTOR LILI TAYLOR: "If there's Facebook, we probably would have at least checked each other's status."

Similarly, in a 2017 episode of Arianna Huffington's *Thrive Global Podcast*, actress Jennifer Aniston, who was on the popular television sitcom *Friends*, stated about a recent conversation she was having with the creators and cast members of the show: "We were jokingly saying that if *Friends* was created today, you would have a coffee shop full of people [who] were just staring into iPhones. There would be no actual episodes or conversations."

That's funny, sad, and true.

I saw it myself in that café on the Upper West Side, and too many other times. Once, on an episode of *The View*, Whoopi Goldberg mentioned that she noticed, during breaks, we were all usually checking our phones. She noted: "I will say it's a little rude, it's a little rude, you know, 'cuz nobody's head is up in a break, nobody's head is up. I just say, I don't need to talk to you all, but it'd be nice occasionally."

She was right. I immediately wondered what real-life moments I had missed because I had been looking at Twitter, texts, or emails during our commercial breaks. I was disgusted with myself.

Then I realized with embarrassment that even I'm not a great friend to my oldest and dearest of friends when I buy into the myth of multitasking.

# #

IT HAD BEEN A LONG DAY. I HAD WRITTEN A COLUMN AND done three live television segments and a few pre-tapes. I didn't get a workout in. A few carrots and some supposedly healthy (but not so healthy) chips had become lunch. It was raining, I couldn't get a taxi, the subways ran slow, and I got home later than I had hoped. I still had piles of work to do, so I took off my TV makeup, ordered some takeout, ate way too fast, and sat down at my computer to complete some research for a show the next day.

As I surfed the Internet, I kept getting caught up in one endless website after another.

Then a few texts rolled in. The social media distractions heated up. My eyes were darting about on multiple screens. And then my phone rang. The name of a dear friend popped up. I answered right away.

"Hello."

"Hey, AJ." Old friend, old nickname.

"Hey."

"What's up?"

"Nothing." Only, that wasn't accurate. I was still scrolling, searching, wandering about the web.

"I'm . . ." And he launched into a story. A good one, I'm sure. Only, I didn't really hear it.

"AJ?"

"What?"

"What else are you doing?"

"Nothing." True and not true.

"I wanted to share that with you, but you're obviously busy doing something else. I'm gonna go." And he hung up.

Jarred by the click, I looked at the phone, my friend's name gone, the screen back to black. I glanced up at the computer and felt sickened by my attention to its ramblings and my inattention to a dear friend. There was no two ways about it. I was fully distracted from life by the content of my machines. The way I saw myself was no longer in line with my behavior. I saw myself as someone who liked to be present from moment to moment, to sit on the stoop and people-watch, to be in the middle of the mountains or countryside with nothing but my favorite person beside me. Here I was, turning into someone too busy on her devices to hear a good friend's story.

I decided that we needed a way to introduce new behavioral norms, to put a damper on the rudeness, inattention, emotional distance, and vacancy that now pass for normal behavior.

Loud FaceTime sessions in restaurants.

Laptops taking up all the available counter space at a coffee shop.

Distracting cell phone conversations on a train or bus.

Diverted glances at a text by caretakers and guardians watching over seniors and kids at a public park.

Dates that look nothing like dates, thanks to small, rectangular, buzzing attendees.

The constant interruption of discussion because of a text or social media alert coming through a friend or colleague's phone.

A guy you had three dates with five years ago messaging you via one of many accessible tech apps while you're in the middle of dinner with your new boyfriend, a reminder of the past you don't want or need in your present.

These are not good things. At least, not for me.

It seems to me that if we admit we have a problem and raise the awareness that many of us are operating in ways that are disrespectful, insensitive, rude, unhelpful to building quality relationships for ourselves—and, in the case of texting and driving, downright dangerous—then we may all want to operate in a better way. Not because someone is telling us to, but of our own volition, for our own individual good and the good of others. Perhaps we'll find that if each of us establishes a new etiquette, a new way of behaving better, we can usher in the next new normal, one that nurtures the people and things we value most.

In other words, we don't need rules or laws or someone else stepping in to regulate this away. We just need *us*. Personal responsibility. We have the self-empowered potential to bring about change and we can self-govern to fix this problem. One by one, and then all together, we can find our way back to civility and intimacy and smelling the roses.

# #

THEN I THOUGHT ABOUT MY QUESTION:
Does anyone make eye contact anymore? Does it matter?

My answer was: *Yes, it matters.* I wrote myself a new mantra: *Eyes Up, Phone Down.*

So simple, yet so important. I needed to FOCUS ON ONE THING AT A TIME, because it matters how and where we give our attention. Life is a string of memories. Do I want to remember what that person said, how they told their stories, what their expressions told me about who they are and what they value—or do I want to miss most of that in order to glance at a buzzing device alerting me that someone, somewhere, has "liked" my photo or tweeted at my link or texted something they'd like me to answer *right now*?

We're missing moments. We're missing smiles and frowns and reflective pauses that teach us about each other. We're distracted—or, at the very least, we appear distracted—present but not fully present, interested but not interested enough, and as a result, the relationships we build are more guarded and shallow, less connected. We end nights out with tech-absorbed friends who are constantly texting someone or Snapchatting a funny moment, feeling like we didn't really get to see them at all. We leave first dates feeling detached and unfulfilled when a phone buzzes during dinner and in the taxi and at the bar. When that happens, you're sharing the night with everyone, so not really sharing the night with anyone. I remember glancing over on a first date once and seeing him peruse a screen full of alerts on his phone after just a couple of hours of hanging out together. He chuckled to himself about who knows what, clicked on a link to something, replied quickly to a text, then exited his private tech world, landed back in the real world beside me, and looked at me like nothing had happened. Only it had. Because he had missed my expression while I was watching him do all of that.

He had missed how it made me feel. If he had seen it, he would've known that the whole episode made me want to run from what relationships were starting to look like in this tech-polluted world.

Even our business relationships are not immune. We exit business meetings feeling less valued and appreciated for our work contributions when a boss is partially engaged in what we're saying and otherwise amused by an incoming email.

Doesn't everyone deserve better?

Imagine the possibilities if we just put the devices away, if we looked at in-person time spent with people as time when the rest of the world wasn't invited to disrupt it. Think of how you'd feel and what you might discover. Give it a shot. Then look back and think about what you would've missed with a small buzzing device sitting between you.

So I committed to putting down my phone and looking people in the eye again. I stopped reaching for the phone in my bag every few minutes when the person across from me would pause to chew their food or take a sip of coffee. Now, when on the phone, I no longer pretend that I can multitask. Because guess what? Something always suffers—either what I'm writing or what I'm hearing or what I'm doing. And when I'm going somewhere specifically to spend time with other people, I do just that—spend time with them, with the people in the room before me, and let the people in my phone wait until I'm done.

I also try not to talk on my phone in public spaces, whether it's a bus, a train, a restaurant, a theater, an art gallery, a museum, a music hall or festival, a library, a gym, a hospital, or a school. (How school even has to be mentioned in this sentence is beyond me. It seems pretty obvious that when kids are in school, they should not be on their phones, and yet it happens all the time.) I remembered a day after work when I was taking some time for myself in a salon, getting a manicure and pedicure, and enjoying the quiet. Then, all of a sudden, the voice of the patron next to me, talking on her cell phone, busted through the space between us and shook me from my reverie. Ten minutes later, I knew everything about her divorce. Fifteen minutes in, I was starting to understand why the guy had left her. By the twenty-minute mark, I knew her husband's favorite color (orange, which she hates), why the cat liked him better, how he insisted on putting BBQ marinade on all the meat, his alluring secretary's Facebook name, how he lied about his promotion, and the fact that, yes, he should get an annual skin cancer check.

In public spaces, it can be maddening to listen to people talk into their devices, especially if the spaces are small, confined, or quiet. So I try not to do it. Remember what happened to the cigarette? It got asked to leave.

We should consider conversations into devices in public spaces to be *the secondhand smoke of the twenty-first century.*

In general, I put my phone away when I'm in public spaces. Think of the visual impact of a room full of people appreciating what's going on around them as opposed to a room full of people with faces buried in smartphones. Just think about it.

OKAY, OKAY, I HEAR YOU, THE PUSHBACK . . .

"That may be okay for you, Jedediah, but I listen to music on my phone, so it has to be out."

What about listening to music on a device that just plays music instead of a device that harbors hundreds of other distractions?

Or it's time you explore a feature called airplane mode, put on a playlist, and toss your phone in your bag or place it in its own spot in the room so that you're not constantly looking at the other things on it. It's a great advantage to no longer have to flip over cassette tapes or change out CDs, to not have to hit fast-forward and impatiently glide over the songs we don't like to get to the ones we do. But that advantage becomes useless, and potentially more problematic than not having it at all, if we are all over the device anyway, replying to this or scrolling through that, delving into a tech wormhole every time. Let the device sit in its place and play its playlist, so you can enjoy the music. *Just* the music. It's doable. I promise.

"But I have to have my phone ready to go, because what if I need to take a picture?"

It used to be that photos were for special occasions. Now we're posing in front of a Whole Foods Market because we had a good hair day. And snapping twenty takes of it to feed our social media neurosis. Sadly, I've done this more times than I care to admit, and it's humiliating on many levels. But I had to stop and ask myself: How about, instead, actually being there in the moment, being present, and enjoying the event, the sky, the park, the sunshine, the time with friends? You want to take a picture? Okay. Let's take it and move on to enjoying the real-life moments. Let's not view the day through the lens of an Instagram filter.

Don't get me wrong, I like Instagram. I like sharing my photos and having them available to look back on. But not when it takes minute after minute away from living real-life moments.

When I'm at a cousin's kid's middle school chorus concert, seeing all the parents there with their phones up in front of them, recording, taking near-constant photos and selfies, texting them, posting them on social media instead of actually listening to and feeling the music, I wonder . . . is anyone ever actually just where they are at the moment, *in* the moment? Do we even know how to do that anymore?

Parents have taken video cameras to events for decades. Back then, the camera was just that—a camera. Whatever happened to good, old-fashioned cameras, with one function, taking pictures or video, no other distractions inhabiting their dimensions, no compulsion-instilling mechanism to leap into a black hole because a text, a Facebook invite, a Twitter tag, popped up while you were snapping that kindergarten graduation photo and now you just *have* to check it?

Let's use our phones in the best way possible. In this case, that means having them with us, but limiting their power so that they don't take over the moment.

"But what if I'm needed in an emergency?"

Which brings me back to the hundreds of years that people existed and survived without cell phones, emergencies included. Remember when Mom or Dad or Grandma would put a little piece of paper on the fridge before leaving the house that said "Emergency contact" with a handwritten number? I remember calling my emergency contact (my aunt) once when my parents were out because I flooded the kitchen with dishwasher soap (which was actually laundry detergent soap) and was knee-deep in suds. That was a real emergency. I called my aunt, she called the restaurant where my parents were eating, Mom came home, and everything turned out okay, no cell phone in sight. But now, because of these little devices that make us constantly and effortlessly accessible to the world, *everything* seems to be an emergency. The cat spit up a hairball? *Bzzzzz.* Emergency. The kids won't stop fighting over which channel to watch? *Bzzzzz.* Emergency. Your sister broke up with her boyfriend and needs to talk to you NOW? *Bzzzzz.* You look down to read her text, miss your kid delivering his one line in the school play—and you hate yourself.

If you feel the need to always be available to some people, consider getting a second phone that exists for emergency calls, with just that feature and no added distractions, no texting or apps or social media or any of that. Give the number to the few people who should have it for emergency reasons—perhaps your child, your spouse, your elderly parent. And set the parameters. Be clear about what it's for. Take only that phone with you to the school play. The other one with the texting, social media, and who knows what else that is eager to interrupt your evening? Click. Off. Believe me, there are too many times I wish I had done just that. Especially that day at a state fair in New Jersey just a few years ago.

## CHAPTER 4

ALL-ACCESS DISTRACTION

Can I tame my Tech Time?

I LOVE THE OLD-TIME FEEL OF COUNTRY CARNIVALS. THE red, gold, and green lights of the Tilt-A-Whirl, the elegant gallop of multicolored horses rounding the carousel, the whimsical twists and turns of the caterpillar roller coaster, the smooth sailing of the *Buccaneer* as it makes your heart race like crazy every time it almost hangs you upside down. Not to mention the aroma of cotton candy and caramel apples, and the *bing-bong, ding-dong* sounds of game winners hammering and ring-tossing their way to a goldfish they'll resent having to feed in two weeks' time.

These things speak to me.

A few summers ago, my friend Sally and I went to the New Jersey State Fair. It was a beautiful night. Perfect weather. Fun crowd of all ages. We were strolling along with no game plan, connecting the way old friends do—little said, much understood. I was negotiating the politics at my job, she at her kids' school, each of us navigating relationships and grappling with the ups and downs of living grown-up lives. The carnival was the perfect antidote to it all.

We made our way past the games and came upon the Ferris

wheel. I love the Ferris wheel. The slow rise into the view over the trees, a chance to grasp a bird's-eye view of miniature humans milling about. It's one of my favorite rides. Sally didn't want to go on, so I ventured alone.

I took my seat in the small silver passenger car, and a smiling attendant double-checked my armrest for safety. The big wheel lurched backward, and suddenly, slowly, surely, we were moving. I giggled and felt that too-rare simple joy sweep over me, my giddy five-year-old self somewhere inside me embracing the feeling of floating up into blue and . . .

Bzzzzz.

Wait, what? Who's there?

Bzzzzz. Bzzzzz.

It was my phone, buzzing with a text. I reacted immediately, almost Pavlovian, and looked down at the number. My heart sank. The name of my ex-boyfriend, the con man you read about earlier who sent me tumbling through a tech nightmare, popped up on the screen. As the Ferris wheel lifted me closer to the clouds, I saw nothing but the white background, angry letters, and well-chosen emojis of my buzzing companion.

I read the text. He wasn't happy. Why should he be? I had broken up with him the day before, after he lied to me. We each needed time to calm down, to get some space, to regroup. Only, now we had these little gadgets of immediate gratification in our hands, so we couldn't quite get that time. One of us didn't have any control over his impulsive need to get the last word. The other of us didn't have the restraint not to answer.

Back and forth we went. Spitting out words of fire across the ether. Head down, brow furrowed, thumbs slamming away at the letters on the tiny keyboard. I was no longer joyful, no longer channeling my five-year-old giddiness, but instead aggravated and annoyed, shouting through the airwaves.

Finally, I had the wherewithal to remember where I was. I was at a small-town carnival, after all, immersed in the daydream I often conjured up for myself when battling the noises and commotion of Manhattan. This was supposed to be my happy place, for crying out loud. Shame on me. I looked up from my phone to take in the pink night sky, when—

Bump, clang.

The ride came to a stop.

I was on the ground. It was over. I'd missed the whole thing. Instead of enjoying the recharge of moving in slow motion and gazing off pensively at the carnival below, memories of my childhood and thoughts about nature or love or humanity organically unfolding before me, I had missed it all.

The attendant opened the door for me and I exited my little car, disgusted. I considered getting back in line to enjoy it again, but I was so irritated with myself that I decided I didn't deserve it. I marched back to my friend, who was leaning against the fun house, caught up in some texting nightmare herself.

Why did I react to the buzz of a message? Then once I saw who it was, why didn't I just put the phone down? Why was I so consumed by the texts, by my ex's near-incoherent rant, by my phone in general?

When I told a friend about it later that night, she joked, "Too bad. Maybe if you'd looked up, you'd have seen someone in the seat behind you who would've helped you forget about that jerk."

I laughed. Maybe I would've. Why couldn't I just enjoy the moment? What was it about technology that was doing this to me? Or better yet, why was I allowing it?

I GOT MY FIRST CELL PHONE IN COLLEGE. IT WAS SIMPLE, for calls only. On it, I had the numbers for my mom, my dad, and a few friends. If someone needed me, they could call it, but they would only do that if they *really* needed me. Otherwise, they would wait until I got home and call me on my landline. In that way, my time and my day were my own. No interruptions other than the ones physically in front of me.

I dated my college boyfriend for two and a half years. I had heard the first name of one of his serious ex-girlfriends in the context of a larger story from one of his friends. I stumbled across a couple of photos of him with an old girlfriend while he was moving and I was helping him clean out his apartment. But the past ended there. His exes didn't pop up on his phone in the middle of our dinners with little heart and kiss emojis and a "Hey babe, wazzz up, I haven't talked to you in a year, what's goin' on?" He didn't receive a request for friendship on Facebook from some girl he dated for five seconds five years ago and in whom he was no longer interested.

None of that happened before cell phones and technology. If someone from the past wanted to come into your life, they had to make an effort to call, to pick up the phone and knowingly interrupt a life they weren't part of anymore, and to risk that your new partner, or child, or some manifestation of your present would answer. Or they'd have to go out of their way to try to find you and see you. Thought went into the whole thing. There was accountability on their end because it would be their voice, their face, that would be met by your present life. You couldn't just cowardly slither your way into another person's consciousness with the push of a button. It took some courage to reenter, and so you didn't do it unless it was important.

Now this constant flood of connections and accessibility can cause as many difficulties as trolling. Anyone who has a phone— and that's almost everyone now—is a target for life interrupted. Say there is a relative with whom you'd rather not be in touch, and you go on Facebook and, because you are both connected to so many of the same people, there they are, in your space. Yes, you can block them, delete them, or tell them to leave you alone. But that's a lot of negative energy you've just absorbed. It's a lot of effort, and you're left with a walk down memory lane that you didn't want or ask for, with a reminder of why you cut them out to begin with stuck in your day. Worse, if you have someone from your past who haunts you in an offensive way, you might find that if you block them, you're giving them the attention and satisfaction they want by validating that they're worth such an action. So you try to ignore them. But they remain aggravating and drive you nuts. Wouldn't it be great if you didn't have to deal with any of that, if you could just be in touch with whom you want to be in touch with, end of story?

I was at the point where I was used to unwanted virtual visitors popping up here and there. An ex-boyfriend on Facebook, a text from someone who didn't treat others well and from whom I had shifted away, a notification that someone who had been toxic in my life had re-followed me on Twitter. Similarly, I found myself acting without thinking sometimes, spontaneously hitting "likes" in people's spaces regardless of our prior relationship complexities, tweeting what I deemed clever comments with little regard for the wording, texting something I'd be unlikely to say quite that way in person, if at all. Just a careless approach to people on the other end of the social media or text portal because I, too, had started to forget that texting, social media, and all of these new web-based interactions deserve the same care, thought, and consideration I bring to my everyday, face-to-face world.

The idea that anyone can say anything to anyone, anytime, affects everyone. It's not always violent and mean. Sometimes it's just invasive and inopportune.

Once, during an episode of *The View* in the spring of 2017, we were on air, filming live, and one of the many phones that were sitting right there on the table rang, right in the middle of the live taping. One of us quickly claimed the phone as hers and turned it off. We laughed about it and continued talking. That ringing phone could've easily been any one of ours who had forgotten to shut off our device that particular day. It was funny in the moment. But I was left thinking: Why are we all okay with constant interruption? Why do we accept that life in 2018 means being bombarded by cell phones ringing or texts or notifications from our devices, telling us that someone, somewhere, needs our attention NOW?

Chances are, you've had this happen. And if those interruptions are casual, maybe you let them go. But what happens when it's more than inconvenient, when it's downright insidious and aggravating? Maybe there is someone out there trying to reengage you via text or social media or LinkedIn even though you've moved on, outgrown them, or decided that you no longer want them in your life. Or maybe a new acquaintance you met in your apartment building's laundry room has started texting you endlessly. Or maybe your work chat app sends messages at 9:00 p.m. that demand attention, or your boss sends emails through the night while you're on the couch watching TV with your wife or eating dinner with your kids. These are boundary issues. These are people who, like the bullies and trolls, can be, while maybe not as threatening, equally as frustrating. You know it. You've been there. Your present personal spaces can now be invaded by everyone and anyone you've ever exchanged emails or phone numbers with in the past—heck, by anyone who knows your first and last name, and can search you on public platforms.

When it comes to work life, so many of us have this problem. You know it. It's constant. Maybe you work nine to five, and then some, because you are always being contacted, via email or the office messaging app, by someone at work. It's familiar, I'm sure . . .

You're sitting at the kitchen counter at 8:00 p.m. The emails start rolling in, then the work app notifications. "Can you fix this spreadsheet?" "Schedule this meeting, please." "Did you modify that email?" "Take a look at this messaging chain before tomorrow." Your boss wants something. Right away. Then again, and again.

There used to be something called the workday. Now it's the

work-never-ends-day. What's worse, often, is that the further up the professional ladder you climb, the less your life is yours. In many work environments, there is the impression that the person who is always accessible has a better chance of getting ahead, of being taken seriously as the ambitious employee worthy of better compensation and professional reward. *If I'm not plugged in all the time, then I won't get promoted* becomes the panicked mantra of the average worker these days. Constant accessibility and availability have become synonymous with upward progression. This goes well beyond the old days of someone staying late at the office now and then, hunched over a desk to finish a project. Because now, even on days when you walk out the door at 5:05 p.m., that office figuratively sits in your hand, in your pocket, then on your kitchen table, or next to your bed, in that little rectangle, insisting that your job take up space in the rest of your life. *Bzzzzz* while you sit with your spouse and watch a movie. *Bzzzzz* as you ask your kids about their school day. *Bzzzzz* as you wind down for sleep.

This needs to shift.

Ask an employee why they're doing so well and they'll say they are working hard.

Ask what working hard means and they'll say they are putting in a lot of hours.

But those hours are no longer confined to a "workday," as the workday now runs through the whole day. Many are even checking work emails when they wake up in the middle of the night for a bathroom run. Talk to a professional in fields like communications, advertising, marketing, public relations, real estate, law, banking, or technology, and many of them will tell you that their availability is their bread and butter—that being "one click away" keeps the clients happy.

Who in the heck trained these bosses and clients to need this immediacy, this constant communication, so much handholding?

In a 2017 article in the Cut called "Please Don't Text Your

Employees at 9 p.m.," writer Dayna Evans reported on the hiring practices of the CEO of a satirical sports and men's lifestyle blog. The CEO said, "Here's something I do: If you're in the process of interviewing with us, I'll text you about something at 9 p.m. or 11 a.m. on a Sunday just to see how fast you'll respond . . . It's not that I'm going to bug you all weekend if you work for me, but I want you to be responsive . . . I think about work all the time. Other people don't have to be working all the time, but I want people who are also always thinking."

Always thinking *about work*?

Always responsive *to work*?

No.

NO.

If work occupies your thought process all the time, it butts into time for family, friends, fun, relaxation, restoration, and other things that complete you as a whole person. Remember that ideal we used to call "work-life balance"? Well, it's dying, if not already dead. As a consequence, your health, sanity, and outside priorities are often suffering. Big-time.

Look at all the good old sitcoms from the 1950s. The dads (TV, at the time, portrayed mostly the dads as the workers) had breakfast with the family, went off to work, came home, and everyone had dinner together. No work in sight. Heck, my dad and mom did that in the 1980s. Barring a real emergency, of which I can remember two, my parents' work didn't have a place at our dinner table or in the evening time after.

A normal twenty-four-hour day back then consisted of four parts, give or take a few hours:

Part 1: the morning, 6:00 a.m. to 9:00 a.m.
Part 2: the work-/school day, 9:00 a.m. to 5:00 p.m.
Part 3: the evening, 5:00 p.m. to 11:00 p.m.
Part 4: the night, 11:00 p.m. to 6:00 a.m.

Now it's all one big blur, no lines drawn in the schedule between work and outside life. The daily calendar has gone flat.

I realize that some jobs may have unconventional hours. Some people have night jobs. Others have jobs that begin before the sun comes up. Everyone's life looks different, and that's okay. What I'm addressing is an endless workday that routinely seeps into your outside life with no sense of *Hey, work, hold up. This isn't your time.* That is what's unfortunately becoming a pretty universal epidemic.

Now combine that never-ending workday with general, never-ending tech overload. I hope the following scenario doesn't resonate with you, but I imagine that for many it will. A typical morning goes like this: you get up, check your phone, and at some point, while getting ready, read and respond to some texts, respond to a few emails, peruse the news and headlines, maybe even click on a few stories. Then you get in your car, drive to work, and find yourself feeling the nudge to check your phone at the stoplights. Or you're on the bus, the train, the subway, and no one is making eye contact because everyone is looking down at their phones. You're listening to music on yours, too, so you start to scroll and quickly fall into a daze of reading old texts, looking at old photos, reading old emails. Now you see a distracting, perhaps annoying, super-urgent email roll in from your boss, about which you can do nothing until you get to work. Then another from the furniture store where you just bought a new couch, suggesting that you purchase some other items as soon as possible, because they are "On Sale Now." Why did you check your phone? You have no idea. But while you're there, if reception allows, you look at your social media. Facebook, Twitter, Instagram, back to Facebook. Before you know it, you have a headache. Your phone has sent you a morning full of headaches. And the day hasn't even officially started yet.

At the office, you spend the entire day bouncing between in-person interactions with the humans at your job, online communications from those same humans at your job, and other humans texting or emailing you on your phone and computer. Emails constantly flood in from the very people sitting right near you all day at work. Endless alerts from a work messaging app that require constant attention and response. Emails from family members, texts from friends. A guy you're dating decides to text-fight about something that transpired on Saturday. On the way out of the office, or on the way home, an associate sends you a message: "One more thing to take care of." Your roommate texts you a list of "things we need for the apartment." By the time you get home, you're overloaded from receiving and responding to one request, communication, and thought after another, all day long. Over and over and over again.

You're exhausted, and little of that stems from the work you accomplished that day. You're stressed. Any wire of energy that wends its way through your body is burnt to a crisp. You never think you're done, because you can't help but wonder, *Did I answer all of my texts and emails? Did I get back to everyone?*

Contrast that to a day in the life of our 1950s sitcom character, or the generations that came before. They sat in the car, on the train, on the bus, and they . . . thought about things. Or had conversations with people. Or read a book. Or wrote a little reminder note to "pick up flowers for the wife" on their commute home. They interacted in person with people at work and occasionally on the phone. It sounds mind-blowingly healthy.

When it comes to work, life suffers when work isn't *part* of your day but *all* of your day. There are only twenty-four hours in a day, so if you're working most of those hours, then you are sacrificing something else: dinner with the family, time with your kids, catching up with an old friend, volunteer work at the community center, conversations and real communication with your husband or wife. If technology enlarges our capacity to do more work, then inherently it has to shrink our capacity in other areas. That is the friction of our lives these days.

No profession is immune. Lawyers are inundated with client requests. Real estate agents have homeowners texting, emailing, and calling with their urgent (and not-so-urgent) complaints. Teachers have parents emailing constantly with questions and concerns. Everyone is accessible. Boundaries are low. How can you hold down the fort when it is constantly under siege?

It used to be that high-stress jobs corresponded with high-stress lifestyles. Those who went to Silicon Valley to work in technology or software start-ups, or to Wall Street to work in banking and finance, were making a conscious choice to work twenty hours a day, six to seven days a week. But now the walls of a regular workday have come tumbling down for jobs that were not supposed to be invaded by technology, but are, with bombardments of emails and texts, day and night. In fact, you'd be hard-pressed to think of someone you know whose full-time job hasn't evolved—and by that you know I mean *devolved*—to become a fixture in the entirety of their days and nights.

When it comes to the rest of your life, real life is compromised by tech's constant access and accessibility to almost everyone, almost everywhere.

The 2017 Pew Research Center statistic that said 95 percent of Americans have cell phones was from an article on the study that also noted, "80% of adults in the United States owned either a desktop or laptop computer; 50% owned tablets; and 20% had some type of e-reader." According to Pew, "Americans are increasingly connected to the world of digital information while 'on the go' via smartphones and other mobile devices."

Everywhere we are, we're connected. It's doing grave damage to our focus and attention. I learned that very quickly when I started teaching.

#

WHEN I WAS IN COLLEGE, I THOUGHT I'D LIKE TO BECOME A professor. I loved the warm energy of my college campus, the old books in the library, the Main Hall's rich architecture that made me feel like I was embarking on an incredible journey every time I stepped inside. Unfortunately, the material I pursued in graduate school didn't resonate with me the way I had imagined it would. I got my master's and moved on, ventured down a few different career paths, and then decided to teach middle school and high school. I taught high school Spanish for one year on Staten Island, then moved to Manhattan and became a ninth-grade academic dean and a teacher of grades seven to twelve at a private school on the Upper East Side. This was about the time that I started to see cell phones everywhere. We had a zero-tolerance policy for phones in the class-room, and confiscating them became a daily event.

I also tutored on the side. Even in these one-on-one, one-hour sessions, I had to deal with a phone constantly buzzing nearby—in a book bag, on the counter, on the kitchen table—so much so that I had to ask students to leave their phones in another room and turn the sound completely off. Otherwise, the second a phone would buzz or light up, they'd stare at it momentarily, then lose all attention in the project or assignment at hand. Additionally, so many students I taught wound up being on medication for attention deficit disorder. I wasn't sure if the increase in this diagnosis correlated with the on-slaught of devices, but I had a suspicion.

In a 2013 article in *Time* magazine's Techland titled "A Nation of Kids with Gadgets and ADHD," Margaret Rock notes, "In the U.S., 6 million children have been diagnosed with ADHD, making it the most common childhood behavioral condition. In fact, over the past decade, the number of kids diagnosed with the disorder surged by over 50%. And in the past six years, that rate has jumped about 15% alone, according to new data from the Centers for Disease Control and Prevention."

The column adds, "The rise in ADHD has coincided with the rise of mobile devices."

Maybe the coinciding is a coincidence. But maybe it's not.

# #

ATTENTION IS A PRECIOUS COMMODITY THESE DAYS. EVERY-one is vying for your attention. Few are getting it. Even you are not getting it from yourself, for yourself, because it is disappearing day by day, spread out superficially over countless platforms, contacts, direct messages, emails, etcetera. The prevalence of OCTD has led to a loss of attention and focus in too many lives, not just those where there is a diagnosed, clinical condition. If we don't pay attention—to ourselves, to each other, to the world around us—or hold on to the ability to do so without distraction, we are going to lose our capacity to be aware, conscious, sentient, thinking humans.

Why are we giving up on our engagement and awareness and, as a result, control of our own lives? For what? Because it's nice and easy to push a button, swipe an arrow, or let computers do the work by shouting for "Siri," "Cortana," or "Alexa"? (Whoa, that's interesting. Notice the gender of the names we order around to assist us. But that's a whole other book.) Is it so important to be on Snapchat continuously, endlessly sending snaps back and forth to others so that we can rack up what they call "streaks" in a competi-tion for nothing except to see how compulsively locked into a task we can be with another person? Our attention is constantly being accessed. And we are letting it. Twitter has us reading our @ col-umn every hour or more. Facebook has us commenting and posting and obsessively sharing videos and reading timelines. Instagram has us hitting likes and checking views and watching stories. When I started realizing how much of my *divided* attention I was giving away to countless sources, I also realized that I was no longer giving *undivided* attention to anyone or anything. All to stay addicted, plugged in, waiting with bated breath and tap-ready fingers for the next feature.

On top of all the other side effects of a tech-addicted life, we are exhausted. In fact, there has been a big to-do lately among doctors and scientists that too many of us are sleep-deprived these days. We go to sleep too late, get up too early, and watch shows or read before bed on electronic devices that emit streams of light that inhibit the melatonin hormone that helps us sleep. We need to sleep.

Yet, we allow our devices, with their melatonin-inhibiting streams of light, to steal our sleep.

This has become such an important issue that in 2017 the Nobel Assembly at Karolinska Institutet awarded Nobel Prizes to three researchers studying our circadian rhythms when we sleep and when we stay awake. The Centers for Disease Control and Prevention, our federal agency that researches and promotes public health, stated back in 2015 that "persons experiencing sleep insufficiency are also more likely to suffer from chronic diseases such as hypertension, diabetes, depression, and obesity, as well as from cancer, increased mortality, and reduced quality of life and productivity. Sleep insufficiency may be caused by broad scale societal factors such as round-the-clock access to technology . . ."

There it is. "Round-the-clock access to technology." A constant compulsion to check our devices. The phone in bed with us while we type on our computer. We type, then check our social media on our phone, then answer a text, back to the computer, the phone buzzes . . . repeat. I swear, it's like these devices see us when we're sleeping and know when we're awake.

We're all working the night shift in this dark digital age. It's killing us.

The cover of the October 15, 2017, edition of the *New York Times Magazine* asked, "Why Are More American Teenagers Than Ever Suffering from Severe Anxiety?" Several examples and reasons were cited, but one, of course, caught my eye: social media. The writer of the piece, Benoit Denizet-Lewis, observed this from a conversation with psychiatrist Stephanie Eken: "Anxious teenagers from all backgrounds are relentlessly comparing themselves with their peers, she said, and the results are almost uniformly distressing."

"Anxious kids certainly existed before Instagram," Denizet-Lewis continued, "but many of the parents I spoke to worried that their kids' digital habits—round-the-clock responding to texts, posting to social media, obsessively following the filtered exploits of peers—were partly to blame for their children's struggles. To my surprise, anxious teenagers tended to agree . . . I listened as a college student went on a philosophical rant about his generation's relationship to social media. 'I don't think we realize how much it's affecting our moods and personalities,' he said. 'Social media is a tool, but it's become this thing that we can't live without but that's making us crazy.'"

Because social media is getting full access to our brains.

MY BIG FEAR IS THAT MANY OF US ARE OKAY THAT OUR time and attention have been hijacked. Too many of us feel it's easier to be on our computers or our phones rather than look each other in the eye and talk—really talk. We prefer convenience and quickness to character-building ways that take some time. We find it draining and hard to engage in person when we can just text instead. Even the idea of a phone call has become an ordeal to many. The dialing, the listening, the inability to do ten things on three devices at once while you talk without making it obvious.

What are we becoming? What have we already become? Think about it. There are two, maybe three, decades of human beings out there who only know a world where small devices control their lives and the lives of their parents. They've been spending most of their lives looking down. They think that's normal.

Bored? Lonely? Itching not to be left out of the loop? There's the answer right there, in your hand. Touch it, tap it, swipe it, and once again you are not alone, you are connected, you are made whole.

I'm sure I'm one of many who would argue that more people than ever feel empty, not whole.

Is that okay with you? I sure hope not.

# #

I DISCOVERED THE ANSWER TO MY QUESTION:
Can I tame my Tech Time?

It was this: *Yes. Create a cell phone curfew.*

The key for me was to approach my tech usage with MIND-FULNESS.

It's easy to be mindless. Once I recognized that, I started to recognize patterns in my behavior and work to change them.

I now block off specific times of the day for work. That includes preparation for appearances on television shows, reading different materials, writing, hosting television shows, and sending and receiving work-related calls and emails. With my job, the schedule can be a little tricky. I'm a good example of someone with an unconventional job with unconventional hours. If I'm working on a morning or afternoon TV show, producers often need to send me stories to peruse briefly at night. So my solution is to designate certain hours of the day for work. In my case right now, they're morning and afternoon hours. In the past, I had to allocate an hour or so at night. But regardless of when I establish those designated "work hours," work doesn't seep into outside time. If someone needs to reach me from work for an emergency, if it's an *actual* emergency, they can do so, and vice versa. But let's be honest, texts and emails after work hours often consist of colleagues who are brain-dumping information, ideas, or tasks from their mental to-do list onto yours. That is, they're just contacting you to *do work stuff*. It can wait. That's why it's called "going to work." That's the time you go, either in person or in the ether, to the place where you *do work stuff.* The rest of the time is, dare I say, yours.

I also realized I am especially prone to Internet clickbait, those attention-grabbing links trying to drive us to some sensationalized story, which gets you talking about the very thing you don't want to be talking about. Once I realized what I was doing, I worked to stop. I'd look at a link, consider if I was going to need or want the information I was going to get there, and then, very often, not click.

When it came to personal devices, I got rid of my iPad, which had become just another thing to check. I currently have one cell phone and a laptop. I also deleted any phone apps that I didn't feel were useful. I turned off social media and email alerts so that I wouldn't constantly be beckoned to those spaces. I could still check my messages and post something when I wanted to, of course, but the onslaught of notifications had become a headache. I kept my phone away from in-person time with the people I love. If I wanted to take a quick photo, I did so, and then put the phone back in my bag. When I played a card game with friends or put on a cleansing face mask and worked on a puzzle, I put the phone in its charging station with the sound off. The world in my phone could wait a little bit while I tended to the world right there in front of me. I started to remember the feeling of uninterrupted time, and I loved it.

I like the information I get online. I like television shows. I enjoy social media. For me, though, it's all about not letting that stuff overtake my real life. I've worked hard to discipline myself, to define my day, to decide what I will and won't do on my computer and phone. I started to own my life again, and it felt great. I remember sitting back and learning to enjoy simple, quiet time again.

The stresses of life make us crave a little distraction. So we engage in activities and entertainment. Traditionally, this was called Leisure Time, maybe spent with family or friends, in conversation, playing games, going for a walk or to the movies. Now, though, that time also seems to be Tech Time, when eyeballs are glued to one device or another, checking and rechecking, scrolling and rescrolling, binge-watching and rewatching. Tech overload has now also saturated what was once our leisure, our escape, our calm. Have we completely forgotten how to wind down, sit still, and enjoy some old-fashioned relaxation without a device in front of us?

Even bath time isn't immune. In 2017, CNN.com revealed: "Teen Electrocuted After Playing on Phone in Bathtub." The fourteen-year-old had just sent a photo to a friend of her phone plugged into an extension cord so she could use it in the bathtub. Though investigators believe the phone never hit the water, they hypothesized that the girl might have touched her (possibly wet) hand to the frayed wires on the extension cord while she was in the tub. CNN noted, "In light of the incident, friends and family have taken to social media to warn of the dangers of texting in the tub."

A devastating story. It used to be that baths were an indulgence, a time of quiet contemplation. In this case, it seemed a secondary thing to do while her eyeballs were occupied elsewhere. Occupying eyeballs. That's actually a mission of many businesses in this day and age. Social media sites certainly want us watching them and each other.

So do entertainment sites.

At the beginning of each calendar year, Netflix sends out its annual communication to its investors and backers. In one of these, a "Letter to Shareholders" dated January 2016, Netflix stated that by the end of 2015, the company had 74,760,000 members. These members, who were all over the world, streamed 42.5 billion hours of entertainment in 2015. If you can average out such a statistic, that's 568 hours per person during that year. That's at least 1.5 hours a day—watching Netflix. Or, more likely, 10.9 hours in a weekend binge-watching some series.

These hours only represent time watching Netflix. They don't include time watching other streaming channels, let alone regular television broadcasts. The 2016 letter stated: "Our focus on 'winning moments of truth' means that we compete with all of the activities that consumers can engage in during their leisure time, such as reading a book, playing videogames, watching linear TV, movie theatre-going, etc. Given the broad array of options, we are privileged that our members around the world continue to devote more time to Netflix, streaming 42.5 billion hours in 2015, up from 29 billion hours in 2014."

Netflix shareholders may be "privileged" by their success, but are we? Think about it in the grand scheme of your life. Add the time you devote to Netflix to the time you give to other online streaming channels, texting, social media, video games, work emails, and other online work-related communications, and constant streams of incoming news online. Even if you don't watch Netflix or don't use some of the aforementioned features, take a minute to think about how much eyeball time in the average day you give to some combination of tech-guided, tech-run apps, spaces, and programs. How many tech-free zones of your life are left? Where do *you* fit in? What tiny corner of your day is even left for that?

I realized that my tiny corner had gotten way too tiny. I needed to restore my brain with a good dose of mindfulness.

The landscape of self-help and rehab facilities was once littered with books, seminars, and centers related to drug, alcohol, gambling, and sex addiction. Now the latest and greatest segment of the population is getting help for—you guessed it—device and Internet addiction. In a 2015 article in the *Atlantic* titled "The Rise of the Internet-Addiction Industry," writer Clare Foran reported on the endless stream of treatment centers popping up to treat tech-related addictions. These centers have detox, support groups, trips into the wilderness, medication—the whole gamut of solutions promising ways to disconnect from devices and reconnect with oneself, with each other, and with life.

All of this immersion in technology is depressing to watch and to experience. Many of the studies I've read seem to be locked in a debate between correlation and causation—that is, those who are depressed tend to use technology more, so it *correlates*, versus the idea that the use of technology *causes* depression. Yet, I have eyes and ears. Just from my observations, I've watched high-tech simulation, engagement, and immersion make people a lot less healthy and fulfilled, both physically and emotionally, a lot lonelier, and a lot more stressed.

It may be too late to nip in the bud our draw to distraction, but perhaps we can dig into the roots and cut off those weeds suffocating our attention span without waiting until it's too late and we have to be sent into a detox or rehab facility. I believe we can and we must. Our dependence on cell phones and technology is behavioral, which means *we* can change our actions and break the cycle of dependence when we want to.

ONE SOLUTION IS TO PRACTICE MINDFUL MEDITATION . . . though, if you're anything like me, you heard the words "mindful meditation" and scrunched your eyebrows, skeptical. I'm not good at meditation. Clearing my mind, being in a state of total calm, and sitting still aren't my strengths. I deeply admire those who have mastered the whole thing, but I'm not there yet, not by a long shot. One thing I have tried successfully, though, is directing myself toward mindfulness while incorporating some helpful elements of meditation. When tech overload starts to get to me, when my computer emails are rolling in nonstop and my text alerts keep lighting up and I'm frantically trying to keep up with it all, I sometimes force myself to turn away from everything for a few minutes and focus on my breath. I know this can sound too simple, but when we're inundated with all this tech pollution, we often forget to *just breathe*. We shove food down our throats at work while we stare at screens because there's no time for a lunch hour. Our work phones are buzzing and our eyes are glued to computers all day long with multiple beckoning apps and features. Our digestion suffers, our concentration suffers, our sanity suffers, *we* suffer. I've found that taking a minute or two to turn away from all things tech, to pause, breathe deeply, imagine something peaceful, and feel connected again with myself, with the real, tactile world and my place in it, makes a surprising bit of difference.

Between grad school and my teaching years, I spent some time waitressing in Manhattan to build up cash while figuring out what my next steps would be. I remember a coworker who used to take a break now and then in a quiet back room to escape the tumult of customers, loud music, and food being sent back over and over because it wasn't cooked quite right. She'd sit in silence, close her eyes, and just breathe. She didn't take out her phone and tend to the noisemakers in there. She took an actual break. She somehow always seemed to be the calmest and most measured of us all, the one least rattled by the chaos within the restaurant. Looking back, I'm sure that those breaks she took and the way she utilized them had something to do with it.

In addition to mindful meditation, I also made some practical changes. I no longer charge my smartphone in the bedroom. Charging it in there used to keep me up way later than I wanted to be. I'd check it periodically before bed and it would fill my head with chaos and distraction that would impede my ability to wind down peacefully from the day. I'd even find myself checking it in the middle of the night sometimes. These were all bad habits I wanted to break. So I did.

I stay off of my computer on weekends. Right now I'm able to confine my work to the weekdays, leaving my weekends computer-free, giving my eyes and brain a rest.

I also don't leave the TV on in the background when I'm home like some kind of Bad News Wallpaper. If I want to listen to a particular show while I cook, for example, great. But a constant, endless flood of noise and information doesn't make me happy. You'd be surprised how healing (although initially shocking) some actual silence can be. Or some great conversations that don't have to compete with background noise.

This was all difficult in the beginning, but after a while I began to appreciate the benefits. It's like discovering parts of your life that were there the whole time, just waiting for you to LOOK UP.

What still bothered me, though, is why I fell into the black hole of technology to begin with, and why it was so hard for me to crawl out. So I dug past the weeds of my distraction, into my roots, and did a little recall into my past. Then I followed this up with some research into what's going on in that mysterious western region of Silicon Valley.

Finally, I figured out what was going on with me.

## CHAPTER 5

THAT ONE-POTATO-CHIP
PROBLEM

How can I stop myself from falling into the
digital abyss?

I WAS NINE YEARS OLD WHEN I FELL FAST AND HARD FOR
Mario. He was cute, mischievous, gutsy, and Italian-American to
boot. I loved Mario, and given how much time we spent together,
I'm sure he loved me, too. He was always available to play and was
one of my best friends in fourth grade. It's because of my relationship with him that I'm fully aware of my capacity to become totally
captivated, losing all sense of time and place, allowing hours and
days to pass, fully engaged, enraptured.

Mario was a reliable companion.

He was also an animated figure inside my Nintendo game console.

As most 1980s kids know all too well, Mario is the lead character of the *Mario Bros.* and *Super Mario Bros.* video games. *Super
Mario Bros.* was the one that first captured my attention as a kid. A
plumber by trade—who can resist a hardworking blue-collar guy?—

Mario and his brother Luigi led players in and out of pipes to secret coin rooms, through underwater mazes while dodging dangerous fish, over flagpoles, and through mini-castles where turtles had to be defeated in order to get to the final castle where Princess Toadstool awaited rescue. To build tension, the levels were timed. Mario couldn't just be savvy. He had to be fast. Even writing about it now, decades later, makes my adrenaline pump up a bit.

I played. And played. And when I was done taking a break for some salt-and-vinegar potato chips or leftover pasta marinara or a snow cone from the ice cream truck, I played some more. I would sometimes invite a friend over and we would play together. I'd be Mario, she'd be Luigi, and we'd team up to battle the Hammer Brothers, carnivorous plants, firing cannons, and whatever came our way.

I was hooked. So were my friends. Our young minds were wired to leap toward the newest craze, be it a video game or music-playing device or fashion trend. My addiction to Mario didn't surprise me.

But my dad's certainly did.

My father is not a tech guy by any stretch of the imagination. He doesn't like gadgets. He listens to college basketball games on an old radio with a turn dial. He likes to watch a good television show, mysteries in particular, but doesn't have a flat-screen TV, Netflix account, or any of that fancy stuff. To this day, Dad still carries maps in his car's glove compartment. On vacations, that's how he navigates around. He goes for long, exploratory walks daily, timing himself on an old-fashioned stopwatch. He still goes to the local bank once a week to make all of his deposits and transactions in person. His home office is lined with his favorite old movies, all on VHS tape, right alongside his favorite books. He reads a hard copy of the newspaper every morning with his coffee and watches an hour or so of news roundup at night. He does have an email account I set up for him a couple of years ago, which he checks twice a week, if that, to read a couple of specialized sports newsletters. He also doesn't have a smartphone. He has an outdated flip phone that he rarely uses. My mom had to beg him to get that for emergencies. The guy has never even sent a text. No joke.

I've asked him about this and he says that he likes his life without all the fuss and intrusion. If he wants to talk to someone, he'll pick up the phone and call them. If he wants to watch a movie, he'll go to the theater or watch one on his television. When he goes for a walk, he likes to look around and enjoy the scenery, saying hello to passing strangers and absorbing his surroundings.

When I look at his life, there is a striking calmness to it all. When he spends time with someone, he focuses on that person, with no buzzing or ringing device nearby. There is a simplicity to the moments, a stillness when he reads a book or sits on the porch and enjoys the sunshine. I envy that. His life is his. He decides who or what gets his attention. And only those invited to the table are present at the table.

Dad makes me nostalgic for a time I can still remember. Which is why, looking back, it's upending to realize that even my father, the keep-life-simple guy with almost no tech accompaniments, almost fell for Mario as hard as I did. And it was all my fault.

# #

GROWING UP, I LIVED WITH MY PARENTS IN A SMALL 2.5-bedroom townhouse on Staten Island. Downstairs was a modest kitchen, bathroom, and cozy living room with a fireplace, couch, and dining table. Upstairs were two small bedrooms, a bathroom, and a home office. The office housed a narrow wooden desk and chair, a few family photos on the walls, two cat-scratch poles so that Scungilli, our feisty rescue cat, didn't claw the downstairs leather couch and hence drive my mother insane, an antique mahogany armoire, and an old television with little dials for channel and volume adjustments that sat in the armoire. It was a small room, but

I loved it so much. It had a window that overlooked trees and the sky. I remember sitting perched up by that window on Christmas Eve awaiting the arrival of Santa's sled and Rudolph the Red-Nosed Reindeer. On the floor, Mom had stacked giant cushy throw pillows up against one wall, and I'd curl up with a blanket, Scungilli by my side, watching morning cartoons or evening family sitcoms. This was the room where my Nintendo also resided, stacked on a small wooden shelf in the armoire beside some of my favorite movies.

When I began playing *Super Mario Bros.*, I'd get lost in a time warp. My mom would shout, "Time for bed!" and I would think, *Didn't we just have dinner?*

Only I had lost the three hours in between dinner and bedtime to Mario.

Once I started, I couldn't stop. If the character would die, I'd start over, zip past the levels I had conquered, and try to defeat the harder boards again. And again. And again. And this is coming from an outdoor kind of kid. Freeze tag, playgrounds, bike rides, kickball, climbing fences and trees . . . that was my mojo. My mom scolded me more than once for exploring construction sites. I skinned my knees almost every week from some crazy outdoor activity I had ventured to try. But suddenly there I was, pushing the outdoor playing I loved so much aside without even realizing it, mission-bound in a virtual, animated world that hopscotch just couldn't compete with.

When you don't have brothers and sisters to corral into your playtime, your parents become your activity buddies. Board games, cards, hide-and-seek—you name it. Mom and Dad played record amounts of *Hungry Hungry Hippos, Sorry!*, and *Candy Land* with me. Mom was definitely off-limits when it came to video games, but Dad seemed like he just might give in. So I begged. And begged. And begged some more. Finally, one day, he gave in.

I remember that night so well. Dad reluctantly sat down beside me in the home office. I gave him the controller, explained the game, and sat back and watched him play. It took him a while to get the hang of it. He grew impatient, tossed the controller, made hilariously frustrated comments, and I'd laugh while watching his Mario get killed over and over by an enemy turtle.

"That turtle made me small again!"

"I need to get flower power!"

Watching your dad in full-fledged fluster while manning a remote-controlled Mario is irreplaceable. After some time focusing in with his singular brand of Dad determination, I saw a change come over him. He was . . . transfixed. He sat up straight and stuck out his chin, his whole body leaning forward toward the screen. His fingers moved slowly, purposefully, then started to twitch faster and faster as he gained awareness of how best to make Mario bounce along from level to level. Then, suddenly—*click*—there it was. His eyes were glazed over. He'd gotten over the learning curve and into the groove, and I knew he was hooked.

We played every night for a week. He'd scream at the TV, rant about his character dying or losing a power, emote passionately to each success and failure. Seeing him react this way, I couldn't believe it. I was honestly beside myself. The rational, practical, stable, panic-free father I knew in reality had turned into a maniac in the virtual world. It was stunning, actually.

One night, he abruptly stopped playing. "This is too frustrating," he announced. "Forget it." He got up and left. I sat with my controller in hand, Scungilli nearby, and thought, *Okay, well, that was fun for a bit.* But then the next night he came in again and started playing with me. Once again, he got all bugged out and frustrated. "Forget this," he said, agitated, as he headed downstairs.

The next day was a Saturday. I had been out playing in the neighborhood with friends. When I came home, I went upstairs and walked past the office. To my surprise, there he was, in the room, playing the game, eyes glued to the television. I asked him what he was doing. He told me to hold on, that he just had to get the turtle and then he would pause it. While en route to the bathroom, I heard him scream, "What the hell? No! That's not right, I jumped over the hammer!"

I turned back to check on him.

"Dad, you okay?" He didn't look up. His face was twisted in a feverish craze I'd never seen.

I went downstairs and talked to my mom.

"He's been playing that thing all afternoon," she said. "He skipped his walk. He's missing the whole day."

My mother called him down to dinner. I'd never seen him eat so fast. I knew he was rushing to get back to the game. I followed him upstairs, sat with him, and watched. A few minutes passed. Then an hour.

"Dad, can I play, too?"

"Wait, I just have to beat this one board," he said, wiping his forehead.

Shocked but sympathetic, I went to another room and watched a movie. Finally, I checked the clock. It was one o'clock in the morning. I told him I was going to sleep. He was still playing the game.

The next day, we met at the kitchen table. He was reading the paper, drinking his coffee. He seemed to be his normal self.

"Morning," I said.

He let the paper drop and looked at me. "I'm not playing that game anymore."

He never did.

# #

MY FATHER FELT THE ADDICTIVE PULL OF THE GAME RIGHT away. At first, he didn't try to stop it. He was drawn in, experiencing something new and exciting. He liked it. It felt harmless enough, a fun escape into an animated world he'd never known. But then he realized, after missing an entire day of his actual life, that something bothered him deeply about that addictive pull away from sunshine and books and dinnertime. He pieced together that those games weren't designed to be played in moderation, but rather to hook the user in, and that there was a price to pay for cycling through the adrenaline rushes that kept the player glued to the screen while real life passed by. He also realized that if someone like him, who was not wired for gaming at all, could find himself in a bizarre mind-set where he was suddenly passing hour upon hour without realizing it, then others more drawn to computers and tech trends might find themselves wasting months, not days. Years, not months.

My father has amazing willpower with pretty much everything, and so he decided to stop playing right then and there. He was so turned off to the whole thing that he never returned to it, even for occasional fun. The impact I saw the game have on him changed the way I looked at it, too. I still played here and there, but was suddenly conscious of how I would feel when I'd get looped in and transfixed in a glazed-over way. I started to hate that feeling, so I'd turn the game off when that would happen and go outside and play. My rebellious chip got activated kind of young, so I decided that if the game wasn't created to be used in moderation, then that was *exactly* how I was going to use it.

Just like that, I was free.

For many years.

Until . . .

In 2016, Nintendo relaunched its classic edition, the one I played as a kid, in a miniature console with thirty built-in old-school games. Jeremy, now my husband but at the time my boyfriend, bought it for me, knowing that (a) it would be a sentimental gift that would remind me of my childhood, and (b) I love miniature versions of life-sized things (dollhouses, Lego sets, tiny forks and spoons . . . I could go on and on and you'd find yourself more and more amused by my insanity). When I opened the gift, my heart smiled and my eyes lit up, not only because it was ridiculously adorable in its tininess, but because it reminded me of the cozy home office I loved so much growing up, Scungilli asleep beside me and my mom's snuggly blanket wrapped around me.

We played the damn game the entire weekend.

I wasn't nine. I was in my thirties, and there I was battling the turtle and cursing at moving fireballs. It was a throwback to my childhood and I enjoyed it. But when I realized, as I was saving the princess, that the day had turned into night, I froze. Uh-oh. Here we go again.

Only . . . not this time.

Jeremy, who had conquered a childhood period of serious video game addiction that made my childhood experiences with gaming seem trivial, looked over at me. His expression said it all. We didn't need to say a word. We knew we had gotten sucked in.

We put the game away and headed out for dinner with friends. We'd play again, for sure, but with our heads on straight and the knowledge that although the games weren't wired for moderation, we now were.

# #

AS ANYONE WHO HAS GONE THROUGH DRUG, ALCOHOL, OR gambling rehabilitation will tell you, you have to stay aware once that little chip of addiction is in you. Many of us have that seed of OCTD that, if provoked, can lead to total immersion in the virtual world and a total loss of time in the real one.

Take the whole binge television-watching phenomenon.

I used to love watching *Beverly Hills, 90210*. As a kid, I watched it the normal way. Once a week I'd get my snacks, sit down alone or with friends, and watch the episode broadcast that evening. Then it was over. I'd think about the show, or talk about it with others, and look forward to setting aside another hour the next week, free from homework and responsibilities, so that I could watch the next episode. That was back in the early 1990s.

Then, in 2013, Paramount Studios released the DVD box set of all ten seasons. I had to have it. I got it for Christmas. I was thrilled. There were 72 discs in the package covering almost 300 episodes, each of which was about 40 minutes long. For anyone keeping track, that's about 200 hours of television. Two hundred hours. There are only 168 hours in a week.

As you might guess, I spent my entire Christmas vacation that year watching, or rather rewatching, the whole series. I was off for two weeks from teaching, and instead of hanging out with friends, taking a vacation, or living some other real-life moments, I was watching the characters of Dylan, Brandon, Kelly, Donna, and the crew live theirs. I'd fall asleep at 3:00 a.m. listening to Dylan and Kelly fight, and wake up at 9:00 a.m. to Brandon's problem-solving.

I watched the whole series. In two weeks.

Then, the self-loathing.

I decided I would do my best to not make binge-watching the norm. I came up with a way to combine the benefits of technology without the costs of the craving to stay immersed. I like a good show, a great series, in its proper place, so when Jeremy and I find a show that we like on a streaming service, we watch an episode. Then we go back to life. Real life. Then maybe we watch another one. We try to pace it out. Of course, we're not perfect. Sometimes we fall into the binge zone. There was a week when AMC's TV series *The Walking Dead* had us mesmerized. But it's not our norm, and when we find ourselves in that zone, we take a step back and mix it up with real life. I'm also not at all opposed to the occasional lazy, rainy Sunday where we grab some snacks and spend all day in bed with a TV series or a bunch of movies. In fact, I love those days. They can be just what you need now and then. What concerns me is a new normal where days upon days, nights upon nights, disappear completely and repeatedly to television series that are released in bulk and obsessively binge-watched (on Netflix, Hulu, or Amazon Prime, for example), with no attention to the outside world—the sun, the moon, the stars, other people—other than checking a nearby cell phone for social media alerts or texts.

Just because content is so readily available doesn't mean we have to binge. Or does it? What is the reason for this season of robots, machines, artificial, virtual everything? If it seems to you that we are drawn into our OCTD, you're right, we are.

And we know why . . .

In 2017, a designer/developer who used to work for Google, Tristan Harris, revealed what has been happening behind the curtain at tech companies and what exactly is making our usage of the devices so addictive. On CBS News' *60 Minutes*, Harris explained that many of the best tech companies specifically design their products to get their consumers hooked on them.

Read that again. Insert any street-corner dealer's product, or the name of a consumer-packaging company that makes snack chips . . . *specifically design their products to get their consumers hooked on them.*

Well, well, well.

Harris said that tech companies manipulate our usage by figuratively slithering into our brains and creating addiction. Brit Mc-Candless summarized the interview on *60 Minutes Overtime*: "That phone in your pocket is like a slot machine. Every time you check it, you're pulling a lever to see if you get a reward. At least that's how former Google product manager Tristan Harris sees it . . . [H]e tells correspondent Anderson Cooper that Silicon Valley programmers are engineering your phone and its apps to make you check them more and more."

One of the scariest revelations in the interview, and there were many, was Harris's comment on the apps and programs that keep teen users "hooked by design," such as "Snapchat's 'streaks' feature [that] shows the number of days in a row that two people have traded photos . . ." And, as if we didn't already know this from our own similar experiences, Harris affirmed that "the anxiety of breaking a streak is real."

When video game and social media developers create their latest and greatest product, their objective is to create a compulsion loop, a virtual cycle in which you are engaged in a process, you have success, the success excites neurons in your brain that release dopamine—which, among other things, can create a sense of satisfaction—and you then have the craving to engage in that loop again. And again. And again. Slot machines, cigarettes, cocaine, potato chips (see previous salt-and-vinegar mentions), physical experiences, and sugar rushes can all release this happy chemical in your brain, providing a feeling of pleasure. As it turns out, it seems that our phones—and not just social media, but the actual phones themselves—do the same thing.

# #

BUT LET'S START AT THE BEGINNING.

If you took psychology in high school or college, you might recall two men, Ivan Pavlov and B. F. Skinner.

Pavlov was a Russian scientist who, in the late 1800s, recognized an idea he called classical conditioning. That is what happens when a subject, let's say a dog, is trained to react to an otherwise neutral object, let's say a bell, because it is repeatedly paired with a stimulus, like a treat. If the dog gets to the point where he salivates when he hears the bell ring, even when there is no treat, that is a Pavlovian response. I have it when I see a short, gray-haired Italian man anywhere near a kitchen. I start craving a good marinara sauce like my poppy used to make—my own Pavlovian response, of sorts.

Pavlov's work informed that of another behavioral psychologist, B. F. Skinner, who theorized that our actions are determined by previous actions, and that behaviors are reinforced or adjusted as a result of consequences. The classic example: you touch a hot stove, you burn your finger, you don't touch a hot stove again.

But what's become even more interesting—and exploitable—is

the fact that these patterns of behavior apply to all of us. Anyone who touches a hot stove burns their finger and learns not to touch it again. We all process that stimulus in the same way. So, for the most part, our behavior falls into predictable patterns. That's why, and how, technology can get very scary very fast. Bear with me.

Eventually Skinner and other behavioral scientists began to recognize that we could motivate and even manipulate behaviors by creating what's called a schedule of reinforcement. This means either continuously or intermittently providing a reward for behaviors in such a way that the reward is contingent on the behavior until the behavior is contingent upon the reward. This is the theory on which our gaming and social media industries depend.

On April 27, 2001, a game designer and PhD candidate in behavioral and brain sciences, John Hopson, wrote an article called "Behavioral Game Design" that ran on Gamasutra.com and has been republished many times since. This article became, and remains, as Hopson himself defined it, "a primer to some of the basic ways people react to different patterns of rewards." It includes psychologically founded recipes for game designers on "how to make players play hard," "how to make players play forever," and "how to make players quit" (or keep them from quitting).

Hopson is one of several behavioral scientists who believe that human reactions and patterns of behavior are not unique and that most people will react in the same way to the same stimulation. When he wrote his article, he took Pavlov and Skinner and applied them directly to the virtual world. Hopson extrapolates on the schedule of reinforcements and talks about what he believes are optimal ratios and intervals, the number of times or the period over which an action should get rewarded to draw in players/users and keep them playing/using. These were grab-hold-and-never-let-go-of-that-user game design patterns he was handing over to anyone who wanted them.

The game designers sure wanted them.

This dynamic was all a big "Wow" to designers because it meant a programmer could come up with one design that would have the same desired effect on so many people. As the consumer product, food, and pharmaceutical industries can tell you, this one-to-the-many strategy is the key to a product's reach and immersion in the market.

Hopson's article became a recipe for software developers. Using his expertise in behavioral psychology and theories of folks like Pavlov and Skinner, and all those who came after them, Hopson revealed a simple but powerful fact—you can make users do what you want them to do. Not only that, you can make them do it over and over again, and never want to stop.

You're with me, right? You see how this applies to our devices and explains why we're using them like opioids? It's your basic compulsion loop. In plain English, folks in Silicon Valley are using behavioral science to create games, apps, and devices right in line with how they need to be structured and programmed to get us addicted.

In a 2014 blog post on his website Gamemakers, Joseph Kim, a game designer and marketing strategist, explains the basis of using the compulsion loop in software design. He notes that inherent in the idea is that one must "believe that human free will does not exist and that the creation of habitual behaviors can be instituted and programmed."

WHAT?

Yes.

I'm telling you this right now: *If you don't know who the robot is, the robot is you.*

Our entire country—the world—is consumed by our devices, our apps, and our social media profiles. We have OCTD. We are in a compulsion loop going round and round, a loop designed by programmers to have us checking in with our social media and our apps—and checking out of real life.

Don't just take my word for it . . .

Time.com published an article in 2015 that revealed: "Researchers in Canada surveyed 2,000 participants and studied the brain activity of 112 others using electroencephalograms (EEGs)." This study, sponsored by Microsoft, found "that since the year 2000 (or about when the mobile revolution began), the average attention span dropped from 12 seconds to eight seconds." The article quoted the report: "Heavy multi-screeners find it difficult to filter out irrelevant stimuli—they're more easily distracted by multiple streams of media," and as a result, the article went on to say, "Microsoft theorized that the changes were a result of the brain's ability to adapt and change itself over time and a weaker attention span may be a side effect of evolving to a mobile Internet."

The writers at Time.com, humorously, or not, noted that "the notoriously ill-focused goldfish" has an average attention span of nine seconds. Microsoft's study showed that "people now generally lose concentration after eight seconds, highlighting the effects of an increasingly digitalized lifestyle on the brain."

That is, the goldfish can focus better than we can.

I'll say it again: WHAT?

SocialMediaToday.com reported on January 4, 2017, "The amount of time people spend on social media is constantly increasing. Teens now spend up to nine hours a day on social platforms, while 30% of all time spent online is now allocated to social media interaction. And the majority of that time is on mobile—60% of social media time spent is facilitated by a mobile device." The article, citing various studies, continued: "Astonishingly, the average person will spend nearly two hours (approximately 116 minutes) on social media every day, which translates to a total of 5 years and 4 months spent over a lifetime . . . Currently, total time spent on social media beats time spent eating and drinking, socializing, and grooming."

That's right, according to that article, people are spending more time tapping away on their phones than hanging out with each other in plain old real life, or even bathing or brushing their teeth.

On May 9, 2017, HackerNoon.com published "How Much Time Do People Spend on Their Mobile Phones in 2017?" Their findings, gathered from various research studies performed by com-Score, Nielsen, Pew Research Center, and others, concluded that the average person spends four hours a day on their phone. Almost half of that time is spent on platforms like YouTube, Facebook, Snapchat, Instagram, and Twitter.

I can't help but think of how many more great athletes, musicians, artists, and thinkers we'd have if the four hours a day spent on these platforms were put into athletics, music, art, or ideas. Those activities, though, take so much work, and the rewards are not doled out in precisely measured, oh-so-satisfying increments. But what do those who spend so much time online get in return for hours and hours spent tumbling further and further into the social media vacuum? You tell me. Ever said to yourself, *What in the heck have I been doing for the last forty-five minutes?* upon landing on an online news article that linked from another website that linked from a Twitter post that linked from a friend's Facebook page?

I have.

There go our precious resources of focus and energy, diverted into a digital abyss.

# #

MY FIRST THOUGHT: *NO WONDER I COULDN'T ENJOY MY ride on the Ferris wheel. One buzz from the device in my pocket, and I had to check it. No wonder I lost my mind when I left my phone in the taxicab. No wonder I feel the consequences of addiction every time I . . .*

And then I had a second thought: *No. This isn't right. I don't have to check anything. No one is forcing me to prioritize a buzzing gadget in my pocket. These are behaviors—my behaviors. This is my life. These are my choices. If I'm going to champion personal responsibility, then I had better live it.*

What was I allowing to happen to me—to my life? How could I change it? Stories like the one on *60 Minutes* helped me to understand why I was experiencing OCTD, the motivation and tactics with which companies were luring me in, and my stimulated leap to respond. But I was angriest at *me*. At *my* choices. As I observed others, I became concerned about them, too, for all of us, for making these choices to allow technology to plunge so deeply into our lives.

Yes, the software developers want to get us in a trance, addicted to apps and the machines that drive them, but here's the key: we have the ability to *not* be controlled. Our minds, willpower, and conviction are stronger than any attempts by Silicon Valley to take them over. Even hypnosis is an *active* act of submission, of letting your alert, conscious self go and allowing others to insinuate themselves into your subconscious. *We* make the choice to put ourselves in that position, to assume passivity, and to let someone else run the show.

When it comes to tech, we have the choice to let it run our lives or not, to passively cycle through the mechanisms Silicon Valley has developed to fuel the addiction *they* want, or to actively reject those mechanisms and set up our own tech boundaries in line with the lives *we* want. Right now, too many of us are blindly walking the path that tech gurus have paved for us without even contemplating what it's doing to our lives. Too many others are aware of the negative consequences of tech addiction but are in a passive state of shoulder-shrugging acceptance.

My bottom line: if you don't care about the consequences of tech overload and tech addiction in your life, that's entirely up to you. I'm a big believer in people building the lives they want for themselves, regardless of what they look like, as long as they don't harm anyone else. But if you do care about tech addiction, what is being lost because of it, and what we still stand to lose, then it's time to stop shrugging those shoulders and do something about it.

Take back your future. Reclaim your life. Be the author of the book of your life, not some prewritten character living out someone else's directives. David Eagleman, who wrote the book *Incognito: The Secret Lives of the Brain*, never thought we had a chance anyway. He wrote, "Almost the entirety of what happens in your mental life is not under your conscious control . . ."

Maybe so. But I still believe in the supreme power of the human mind and our power to act in our own best interests. I believe in you and me.

# #

I DISCOVERED THE ANSWER TO MY QUESTION:
How can I stop myself from falling into the digital abyss?

It was this: *Switch off the Silicon Valley Slot Machine.*

I began to use my technology in MODERATION. This approach resonated with me, as I know I'm not, and don't want to be, a cold-turkey kind of gal. My dad put that Nintendo controller down one night in the 1990s and never picked it up again. That's not me. Instead, I embraced the idea of a life in which video games, social media, content streaming, and other tech components would have a role, but that role would not supersede real-life moments, and that role would be decided by *me*. My presence on social media and my phone usage have both significantly decreased while researching, thinking about, and writing this book, thanks exclusively to my motivation to make those changes. I promised myself that whatever compulsion loop Silicon Valley was hoping I'd hop on—well, they could count me out. As I've said before, I believe in people. Because of that, I can't—I *won't*—sit idly by and watch these machines make machines of us.

Surprisingly, some tech executives are on the same page.

In the *Guardian* on October 6, 2017, an article by Paul Lewis titled "'Our Minds Can Be Hijacked': The Tech Insiders Who

Fear a Smartphone Dystopia," begins with this: "Justin Rosenstein had tweaked his laptop's operating system to block Reddit, banned himself from Snapchat, which he compares to heroin, and imposed limits on his use of Facebook. But even that wasn't enough. In August, the 34-year-old tech executive took a more radical step to restrict his use of social media and other addictive technologies. Rosenstein purchased a new iPhone and instructed his assistant to set up a parental-control feature to prevent him from downloading any apps."

A tech executive significantly restricting his tech usage. The article continues:

"He was particularly aware of the allure of Facebook 'likes,' which he describes as 'bright dings of pseudo-pleasure' that can be as hollow as they are seductive. And Rosenstein should know: he was the Facebook engineer who created the 'like' button in the first place."

The guy who created the Facebook "like" button is deeply concerned about . . . the Facebook "like" button? The story goes on to reveal:

"A decade after he stayed up all night coding a prototype of what was then called an 'awesome' button, Rosenstein belongs to a small but growing band of Silicon Valley heretics who complain about the rise of the so-called 'attention economy': an internet shaped around the demands of an advertising economy."

Serious tech concerns coming from right inside Silicon Valley. The article goes on to say:

"These refuseniks are rarely founders or chief executives, who have little incentive to deviate from the mantra that their companies are making the world a better place. Instead, they tend to have worked a rung or two down the corporate ladder: designers, engineers and product managers who, like Rosenstein, several years ago put in place the building blocks of a digital world from which they are now trying to disentangle themselves."

They built it, they see what it's doing to people, and they're limiting its presence in their own lives. More from this report:

"'It is very common,' Rosenstein says, 'for humans to develop things with the best of intentions and for them to have unintended, negative consequences.'"

Negative consequences? Ya don't say.

The column's author, Paul Lewis, adds, "There is growing concern that as well as addicting users, technology is contributing toward so-called 'continuous partial attention,' severely limiting people's ability to focus, and possibly lowering IQ."

I don't know about you, but I'm not in the market for getting dumber.

Smart phone, dumb me?

No, I don't think so.

WHEN I WAS IN COLLEGE, I TOOK A CLASS CALLED EXISTENtial Psychology. The idea of this course was that a person faces issues as a result of merely existing and being alive, of having choices, making decisions, and being accountable and responsible for those decisions. Our brains try to create order out of chaos, we search for meaning, we confront the idea of not existing—i.e., dying—and all of it can be very trying for our psyches.

Dr. Miles Groth, a professor of psychology and the author of *After Psychotherapy: Essays and Thoughts on Existential Therapy*, who taught the Existential Psychology class I took in college, recently shared this with me: "What existential philosophy teaches is that people are free. And they've begun to forget that, to lose sight of the fact that they can make choices for which they have to be responsible. No matter what those choices involve, freedom implies responsibility, a responsibility to and for yourself, to build the life you want."

Writer Jane Spear, on the JRank Psychology Encyclopedia website, sums up how thinkers like Dr. Groth came to the existential school of thought. In the seventeenth century, philosopher and mathematician René Descartes said, in Latin, *"Cogito ergo sum,"* which translates to "I think, therefore I am." Then in the nineteenth century, philosopher Søren Kierkegaard flipped this on its head and said, "I exist, therefore I think." This then got the philosophers and the psychologists really thinking, and existing, and thinking again. (See why philosophy and psych majors get headaches?)

Later on, in the late 1800s and early 1900s, the American educator, scholar, writer, physician, and religious philosopher William James, who is best known for his works *The Principles of Psychology* and *The Varieties of Religious Experience: A Study in Human Nature*, introduced Americans to the idea of pragmatism, personal responsibility, and free will. In the mid-1900s, psychologist Abraham Maslow drew up his famous pyramid of priorities for living, beginning with, at the basic level, the need for food, shelter, and security, and rising up to the higher levels of existence like relationships, love, fulfillment, self-esteem, and self-actualization, i.e., being the best you can be, or even a little better than that. In the mid-1900s, writers Jean-Paul Sartre and Simone de Beauvoir pushed the ideas of free will and individual responsibility into modern culture with their books and plays on the subjects of existing, of thinking, of choices, of being human.

The thread that runs through all this, through Kierkegaard, James, Maslow, Sartre, de Beauvoir, Dr. Groth, and dozens of other philosophers, is the idea that life is both difficult and rewarding. We're faced with hurdles and obstacles through which we must struggle to survive, and if we are some of the fortunate few, to thrive. Our place in the world is what we make it, and the journey is ours to take, to overcome. It's personal, it's physical, it's mental, it's spiritual. There is work and thinking to be done, and when we do it, we do better. The gist, as I see it, of existential philosophy is that we must really understand what it is to be in the lives we are living in order to fully live the lives we are living.

Though that sure is difficult to do if I start to think that everyone else is living a better life than I am . . .

CHAPTER 6

PERCEPTION DECEPTION

What—and who—is real?

ONE EVENING, I WAS WALKING ALONG THE BROOKLYN
Heights Promenade, considered one of the most romantic spots in
New York City, the perfect place to view the one-of-a-kind stream-
lined geometry of the buildings that crowd Lower Manhattan and
the excitement of the boating activity on the East River, an enticing
getaway to reflect and rejuvenate. This was exactly what I needed
so desperately as I made my way to meet an old friend. This should
have been a meditative stroll, sentimentality seeping in as the day
turned to night, the perfect relaxing preamble to a fun evening. The
sun started to sink into the Manhattan skyline, and so I took a photo
of it. Sunset over the city, perfect for an Instagram post.

Only, it wasn't. I looked at the photo and thought, *Oh, this isn't
good.*

And so I took another.

Then another.

Fifteen minutes later, the sun was set and gone. I sat down on a
bench to view all of these Stages of the Real Thing on my phone.
I looked them over and over again until I finally chose one I liked.

Then I put in a filter.

No, not quite right.

Then I tried it without a filter.

That would do it. Good.

I posted it. #NoFilter.

An older gentleman passed. "Such a gorgeous sunset, isn't it?"

I was about to say yes, and then I realized, I didn't really know. Because instead of standing on the edge of the boardwalk, making a wish, filled with the anticipation of seeing a friend I hadn't sat across from in eight years, recalling similar footsteps I had taken along that same promenade when I was eighteen years old and about to embark on my first big date, I was busy collapsing that beautiful real-life moment into an Instagram post. In fact, the only version of a sunset I saw that night was through the lens of my smartphone, through a small rectangle whose unfiltered setting still had the filter of technology over the real deal.

I was embarrassed. I had been lucky enough to have that beautiful sight before me, the chance to think and reflect on a painting in the sky, and I wasn't even really there. Heck, I could've stayed home and looked up "sunset" on my computer and seen the same thing.

#MissedIt.

EVERYONE'S LIFE SEEMS BETTER ON INSTAGRAM. I KNOW that. You know that. Yet, we all look at Instagram, at the #SoBlessed, #LoveMyLife, #BestKidsEver hashtags, and wonder why we don't look like the woman in the photo holding the perfect baby on the perfect porch swing on some perfect lake.

When you scroll through Instagram or Facebook, you are supposedly watching someone else live. Only, you're not. You're watching the moment someone captured, real or manufactured, and decided to show you. It might be edited, photoshopped, filtered. It might even be scripted or completely phony. You'll likely never know.

Social media is creating two worlds—the one that we often think is real, that we are watching unfold in that space—and the other one, actual real life, which often doesn't resemble someone's social media presence at all. We often find ourselves believing people's social media stories and the way they portray themselves and their lives there, without even thinking.

Take it from me: perception deception is real.

I remember a woman on Instagram who posted a photo of herself with her husband that looked like it was right out of a *How to Have the Best Marriage Ever* handbook. She even hashtagged it something about Mickey and Minnie Mouse. Don't ask. The thing was, though, I knew that woman. She spent most of her real-life days complaining about her husband and threatening divorce.

Another time, I was scrolling through my Instagram feed and a picture of a guy I knew popped up. In the photo, his stomach was clenched within an inch of his life and the light was coming through the window perfectly to highlight six-pack abs. There he was, a Ken Doll hunk emanating off my feed. The thing is, I had seen that guy a week before in the gym, right there in person, and he didn't even have the illusion of abs. It was a manufactured moment.

As annoying as this is for us, the adults, it's worse for kids. Boys and girls are growing into their bodies. It's an awkward time. I can't imagine how undermining it is for them to take measure of their life, their looks, their changing skin and teeth and weight, by comparing themselves to the often fake, filtered, photoshopped lives they see on their iPhones. Sure, the images of airbrushed models in fancy magazines have messed with kids' heads for decades, but nowadays there are app-enhanced images on phones making kids think that they're looking at regular girls or guys just living life—when they're not. If and when they find out, they may feel the need to compete by using apps to remodel *their* faces and bodies in images, which may lead to them being altogether unhappy with themselves in real life.

Let's not forget the competition to see who can look better with the most outrageous, scandalous photos: "Well, she is a quarter naked, so I should be half naked." You find yourself doing the most outrageous things.

At one point, I was doing a lot of squats in the gym and actually took gym photos that highlighted my butt. How humiliating is that? It's awful, and I know that some people will have a field day with this information, but it's more important to me that I speak from a place of having made some of these ridiculous mistakes. I am as guilty as the rest. I remember taking a photo of myself looking at a bed of roses as if I'd just come upon it, when in fact, I'd primped and processed the whole shot. The real story? I didn't even really see that flower bed, smell the flowers, or reflect. The flower would comment, if it could: "This is bullshit, you didn't even really look at me."

SOFTWARE DEVELOPERS USE THE SAME METHODS TO HOOK us on video games that they use to hook us on vacation pictures and political memes. Yet, while many of us are aware that software developers and tech manufacturers are reading our minds and manipulating our behaviors, somehow, amazingly, we accept it.

In an article in the *Australian*, reporter Darren Davidson wrote: "Facebook is using sophisticated algorithms to identify and exploit Australians as young as 14, by allowing advertisers to target them at their most vulnerable, including when they feel 'worthless' and 'insecure,' secret internal documents reveal. A 23-page Facebook document seen by *The Australian* marked 'Confidential: Internal Only' and dated 2017, outlines how the social network can target 'moments when young people need a confidence boost' in pinpoint detail. By monitoring posts, pictures, interactions and internet activity in real-time, Facebook can work out when young people feel 'stressed,' 'defeated,' 'overwhelmed,' 'anxious,' 'nervous,' 'stupid,' 'silly,' 'useless,' and a 'failure,' the document states. After being contacted by *The Australian*, Facebook issued an apology, and said it had opened an investigation, admitting it was wrong to target young children in this way."

Okay, so Facebook can allow advertisers to target users based on monitored, real-time emotions. Meaning Facebook now knows how you feel. And the company puts advertisements on their pages that will take advantage of those emotions. So one minute you're feeling the blues because of winter's short days, and the next thing you know, a link to a vacation in Bermuda pops up. But—ready for this? Even before we discovered this—in fact, three years before, in June of 2014—it was revealed that Facebook did something much more underhanded: it secretly studied and *manipulated the emotions* of 689,000 of its users.

In his 2014 article "Facebook Reveals News Feed Experiment to Control Emotions," the *Guardian*'s Robert Booth put it this way: "It already knows whether you are single or dating, the first school you went to and whether you like or loathe Justin Bieber. But now Facebook, the world's biggest social networking site, is facing a storm of protest after it revealed it had discovered how to make users feel happier or sadder with a few computer key strokes. It has published details of a vast experiment in which it manipulated information posted on 689,000 users' home pages and found it could make people feel more positive or negative through a process of 'emotional contagion.'"

This idea of emotional contagion means that other people's feelings can infect our own, if we allow ourselves to fall into unconsciously copying the emotional expressions of those around us. Which, apparently, we do.

How does Facebook spread this contagion?

An article on the same topic by Robinson Meyer titled "Everything We Know About Facebook's Secret Mood Manipulation

Experiment," published in 2014 in the *Atlantic*, revealed: "Some people were shown content with a preponderance of happy and positive words; some were shown content analyzed as sadder than average. And when the week was over, these manipulated users were more likely to post either especially positive or negative words themselves."

Meyer went on to share that Facebook said that it did, in fact, take user data, analyze it, and test out some ideas and processes on its users. Some of those outside of Facebook whom Meyer interviewed cited that, although it may not have been ethical, it was legal. Many dismissed the study as minimal and harmless until they really thought about it, which is exactly my point. When are we going to stop and think about what is going on? As Susan Fiske, the Princeton University psychology professor who edited the study for publication, told Meyer for the *Atlantic* article: "It's ethically okay from the regulations perspective, but ethics are kind of social decisions. There's not an absolute answer. And so the level of outrage that appears to be happening suggests that maybe it shouldn't have been done . . . I'm still thinking about it and I'm a little creeped out, too."

Ya think?

Is Facebook conducting these studies just for . . . fun?

Ha. Ha, ha, ha . . .

Okay, in case you can't tell, that's me laughing in a nervous, holy-cannoli, end-of-the-world kind of way. This happened in 2012. Some were "creeped out" by it, and yet, here I am sharing it in 2018—and no one seems to care. Why? Because we are so addicted to social media that many of us need the smallest nod of "Don't worry about it" to not worry about it. Any of it. Including the fact that software manufacturers not only know plenty about what we like and dislike—heck, we hit "like" buttons on posts, comments, and photos on many social media apps—but they also may be manipulating what we feel and do. There we are, surfing around Facebook, falling into the rabbit hole that is a college roommate's posts, and all of a sudden we're sad, or angry, or depressed. How did that happen? We don't know. But we'd better care.

What if the *Australian* hadn't contacted Facebook? Would they still be manipulating users? Are they still manipulating *us*? Are other apps doing the same? Are they changing how we treat people, what we value, the status of our physical and mental health, or how we are reacting, physically and emotionally?

In the October 15, 2017, issue of *American Family Physician*, the peer-reviewed journal for the American Academy of Family Physicians, Drs. Kaitlyn Watson and David C. Slawson wrote an article called "Social Media Use and Mood Disorders: When Is It Time to Unplug?" The findings begin with a doctor's case scenario: "A 25-year-old woman presented to my clinic with some mood issues that she was experiencing. From the moment I greeted her, she was entirely engrossed in her smartphone, rarely taking her eyes off of it. I could see it was the Facebook app that was demanding all of her attention. The patient described feelings of depression and anhedonia [*I looked this up for the non-doctors and non–Latin majors among us, and anhedonia is the inability to experience pleasure in activities that were once enjoyable*], as well as difficulty sleeping and concentrating. When I asked what was going on with her friends on Facebook, she broke down in tears, explaining that several of her friends had recently gotten married, one had a new baby, and some were even working abroad. Meanwhile, she had yet to find a fulfilling job, was not in any kind of serious relationship, and could not afford to travel. She was happy for her friends but was also constantly reminded of the seemingly fabulous lives they were leading every time she looked at social media. She could not help but compare their lives and experiences with her own."

Such patterns of moodiness, depression, anger, and other outward expressions of inner turmoil worry me. I can see how, in this time of tech overload where we disconnect from real-life moments and real-life people more and more, it becomes too easy to also disconnect from our sense of peace, calm, optimism, and kindness to ourselves and others.

The case study was followed by commentary: "Americans spend more time on Facebook, the world's largest online social network, than any other website. On the surface, social media networks provide an 'invaluable resource for fulfilling the basic human need for social connection.' However, rather than enhancing well-being by fulfilling communication needs that are deeply human, current research suggests that these online platforms may undermine it. Much of the current social media literature has focused on social media use and the fear of missing out, or FOMO, in the millennial age group (typically defined as persons born between 1980 and the early 2000s), although some research suggests that it is not limited solely to millennials."

We've heard stuff like this before, discussions about how social media can negatively affect someone's mood and well-being. But this is in a journal for general practitioners and family doctors because it is relevant to these primary caregivers. Why is it relevant? Because they are seeing more and more patients who are depressed, moody, and feeling downright hopeless as a result of too much social media.

Ironically, the column also included a table of relevant apps to help people use their apps less. Come to think of it, maybe that's not ironic at all. It just is. We are buried so deep in our tech addictions that many can only suggest tech to get us . . . out of tech.

I started to think further about the ability of social media to change our perception of the outside world and ourselves. And then I got to thinking about technology's impact on our perception in general.

MY MOM WAS ALWAYS TERRIFIED OF MOST AMUSEMENT park rides, so it was my dad who ventured into the world of roller coasters, freefalls, and all sorts of adrenaline rushers with me. Every year, Dad would take a few of my closest middle school friends and me to Six Flags Great Adventure in Jackson, New Jersey. He would load up the car with snacks, and we'd play road trip games on our way there, then spend the whole day hitting the rides. He went on every roller coaster with me, even the ones that some of my friends were too scared to try. Dad and I explored local carnivals in New York and New Jersey, Dorney Park & Wildwater Kingdom in Allentown, Pennsylvania, and—when we went on our annual vacation to Florida—he would include side trips to Busch Gardens in Tampa, Disney World in Orlando, the Boardwalk and Baseball theme park in Haines City, and others.

Before I got married, I told my mom that if I could exchange my wedding vows at the top of a roller coaster ride, right before the first big drop that makes your heart race, I'd love it.

Mom looked up from the book she was reading and said, "Why do you drive your mother crazy?"

Okay, maybe I didn't need a roller coaster wedding.

Instead, I decided to take an amusement park vacation with Jeremy. As it turned out, the producers of *The View*—which, at the time, I was cohosting—had planned to take us on a road trip to do the show live from Disney World in Orlando. It was perfect. Jeremy loves amusement parks as much as I do, so we decided to make a vacation of the whole thing.

Disney was amazing. Each day, after taping the show, Jeremy and I would hit the parks, go on all the great rides until we got hungry, then head over to Epcot's World Showcase restaurants, a circular escape that features food from around the world. On our last day before heading to Fort Myers for the weekend to spend some do-nothing time on the beach, we decided to hop over to the Universal Studios theme parks. Universal was known for those three-dimensional, virtual-reality worlds that people seem to love so much. We were eager to see what all the fuss was about.

First, we went on Transformers: The Ride—3D. We donned
3-D glasses, got in a car that ran along a track, and away we went—
up, down, sideways, backways. Water pipes broke, battles ensued,
missiles exploded. Our eyes darted about, we gasped. Water sprayed,
hot air and fog appeared. Every nerve was activated, our jaws and
fists clenched in anticipation of the next invitingly surreal moment.
We were exhausted by the time the ride came to a halt.

We exited, craving more, running toward the next virtual experience. It had to be The Amazing Adventures of Spider-Man.
Spider-Man has always been my favorite superhero, so I was ready
to roll. We strolled through the ride's introduction, where, through
video and set designs, we were prepared to accept our role as crime-
fighting reporters in the narrative. Then we put on 3-D glasses,
complete with night vision, and got into the cars that rode along the
tracks and into the experience. Immediately, we were fully immersed
in the sights, sounds, and vibrations of villains breaking through
walls and floating around on hovercrafts, of lasers shot across our car
and fireballs thrown right at us. We even got wet from water spray
and warm from blasts of heat. Our eyes didn't know where to look
first. Our senses were on overload and we were loving it. It felt like
we were flying through the air, spinning in space, falling fast past
city skyscrapers. Our hearts were racing, our skin tingling. I had to
catch my breath a couple of times. It was thrilling.

I later learned that the tracks were in fact flat. I'd never gone up
and I'd never gone down. But it was more lifelike than any roller
coaster I had ever been on.

Both rides were mind-blowing.

Transformers landing on our car. Spider-Man shooting webs and
flying past us. It was fantastic, and we were lost in it. We had entered
a whole other realm. No doubt, they were two of the best rides I had
ever been on.

We were exhilarated and couldn't stop talking about what we
had just experienced as we walked to The Incredible Hulk Coaster,

a ride that loomed ahead, its green tracks promising high speed and twisting, turning, upside-down fun and thrills. We hopped on the line and waited for the front car. Excited, we looked at each other as it started.

And then . . . it was done.

"That was it?" I said as we got out. "I mean, it was okay, but . . ."

"I feel like it was missing something," he said.

"Yeah, it just felt kind of simple, just loops and a little boring, nothing like the other . . ."

We both went wide-eyed.

Uh-oh.

We were thinking the same thing.

The virtual world had upstaged the real one.

The Transformers had transformed us.

We had spent an entire day at Six Flags Great Adventure the summer before, on ride after ride like the Hulk coaster, and had never felt bored. But now we had experienced two rides on a heightened new virtual level. How could a roller coaster compete with a whole 3-D experience?

I should've realized trouble was afoot the moment I heard the Transformer ride's video introduction, a warning from their leader, Optimus Prime, "Humans, you are in grave danger . . . I must warn you that at this very moment, the Decepticons are mobilizing to take over Orlando . . . their plan is to destroy the human race."

Bothered by our unimpressed reaction to the roller coaster, we sat on a bench in a small cove to debrief.

ME: "When you have the glasses on in the 3-D rides and feel like you're flying and have the wind on your face and your hair blows back, and you see waterfalls that look like waterfalls, and feel the spray . . ."

JEREMY: "It feels so real. And you're taken in. It's like your eyes are bugged out and you can't even process it all."

ME: "People are going to be desensitized by all this."

JEREMY: "To basic rides, to plain old life."

ME: "How do you take your kids to the Grand Canyon or Niagara Falls and expect a *WOW* when they've seen all this manufactured virtual stuff where every one of your senses, your nerve endings, feel like they're on fire?"

Beams of orange sunlight radiated through the hedges bordering the cove. We both looked up to see the sun settle slowly into the horizon, and Jeremy squeezed my hand.

"That's still something to see," he said.

I was grateful. Grateful for his hand in mine, for the moment together outside, for the fact that we both could still sit on a park bench and be wooed by a sunset. That was what life was about. A 3-D ride, though fun and definitely an experience, was in its proper place of being just that, a ride. It still couldn't compare to our mysterious earth circling a glowing star. At least, not to us.

# #

BUT WOULD OTHERS SEE IT THAT WAY? WHAT ABOUT KIDS growing up in a world where 3-D this and virtual that are pretty much at their fingertips? Would that be their perception of reality?

That night at dinner, Jeremy and I talked about the virtual world—the virtual, robotic, computer-aided, simulated realities we have allowed into our lives, our work, and our homes. Robot vacuums cleaning your living room floor. Robot systems at your job organizing most forms of internal and external communication. Robot computers managing food and drink orders at bars and restaurants, playing your music at the gym, controlling the lights and temperature in your home, securing the doors and windows. Robot-run apps and social media platforms. Virtual reality and 3-D computer games. Alexa's robotic voice responding to your commands to order this item or dim that light or tell you the weather wherever, whenever. Alexa, who's always listening.

It's all happening. It's there for the taking. There for taking over.

Our virtual reality experience at the amusement park took Jeremy back to his time with video games as a kid and teenager. He had played some compulsively. He had used them as a crutch, he said, to not have to deal with some of the challenges and hurdles of real life. Instead of facing those difficulties head-on, be they awkward middle school conversations with girls that help the social transition to high school or learning the lesson of naturally outgrowing some friends and making new ones, he would just escape into the games, absorbing himself in animated scenery, fake characters, and on-screen battles. The problem with that, he said, was that he also missed the growth that came from facing those real-life challenges head-on. Even his appreciation for the plain old outside world, which didn't come enhanced with the vivid, ornate scenery and animated figures of the games, had to be reset. He emerged years after those intensive gaming days having to make up for lost time in the real world in many ways. He still plays the games once in a while for fun, but with a very different eye and attitude, one that manages the technology instead of letting the technology manage him. A few times I did see him catch himself, though, after a few lost hours absorbed in the screen, remembering many lost hours before, and he'd take a nice, long break from the games when that happened.

Jeremy and I have both found, as we did on the virtual rides, that we can be absorbed by the pixels and fabrications of computer-generated fictions. I'm concerned that with more easily accessible exposure to heightened virtual this and that, we will all be less able to appreciate what's real. What if one day we can no longer appreciate a garden pathway, a pine grove, or Yosemite? I've been to places, gorgeous beaches, where I've heard someone next to me say: "It's pretty. It looks just like my screen saver."

Oh, boy.

I then started thinking more about online video games and how a different kind of perception deception had caught a few people I knew by surprise.

A friend of mine, let's call him Jimmy, used to like to play a multiplayer online role-playing game on his computer called *City of Heroes*. In the game, Jimmy interacted with other live players, all of whom had created characters with superpowers who could work together to fight crime. Eventually, Jimmy got into a dialogue with one of the other players. Over time, the conversations got less about superheroes and more personal, with some details that made Jimmy uncomfortable. The player was confiding in him. Often, he wanted to talk rather than play the game. He told Jimmy that he was dealing with major health issues and that he was in a bad and abusive relationship. When Jimmy asked more about it, the player finally said, "I'm not who I said I was." Jimmy found out that the other player wasn't a man around his age, as the player had originally told him, but a woman twice his age, in her sixties.

When I taught high school, one of my students, who was fourteen, had a video-gaming friend he talked to online after school. As they continued to engage more and more, his mother became concerned and started looking into the identity of this gaming friend. Together, she and her son found out that the other kid wasn't a kid at all, but a thirty-year-old man playing out of his apartment. She shut that down fast.

# #

THESE STORIES WERE DEEPLY CONCERNING. I COULDN'T stop thinking about the virtual world, the real world, and everything in between.

In doing so, I had discovered the answer to my question: *What— and who—is real?*

It was this: *To some extent, and according to many philosophers, we decide what is real.* I take the virtual for what it is, am cautious of the information I find on the Internet, and, for crying out loud, have stopped looking over my neighbors' Instagram fences. *The grass isn't greener. It just has a filter on it.*

It was time in my tech life for reflection. I needed to PAUSE AND TAKE CHARGE, and to make some good choices to help me feel better about how I use social media and the technology in my life.

I like virtual rides. I'm amazed by the places technology can take my imagination. As a day here or there at an amusement park or playing a new, fun game, I think it's okay. I get the appeal. But I refuse to allow that life to become *my* life. I won't allow myself to become immune to the beauty and realness of the world, in the same way that I won't allow text messages and apps and social media to have a seat at my dinner table. I support the ability of new ideas and inventions to have the chance to flourish in a free market, but I also understand that at some point we need to think about what we embrace and encourage to prosper, what the consequences are, and what we value. What we pay for succeeds. What we buy thrives. That's a big responsibility, but it's also an incredibly empowering one.

Right now, people are paying for virtual reality, and so it is thriving and, in some ways, winning over the real world. There are countless examples of people spending more time on their computers, on their phones, on their tablets, than they are in reality. How many kids would trade in their handheld devices for some distraction-free time outside? As the years pass, are the numbers getting smaller?

For me, a big step was to pause before any embrace of new technology, be it a purchase or a download, to consider if and why I'd need the latest, supposedly greatest tech innovation. A new app that everyone is using, a social media site that's been labeled the coolest ever, a hot-off-the-3-D-printer gadget that's on everyone's holiday list? I don't have to buy it, be part of it, own it, or advocate it—

unless I want to. I'd also consider whether or not it would improve my quality of life and the impact it would have on my time, my relationships, and my stress level. These are the things that now go into my equation for purchase. Its newness, and how many of the cool kids are buying it, does not. As behavioral scientist Robert Cialdini explained in his seminal 1984 book, *Influence: The Psychology of Persuasion*, one of his six weapons to get other people to do what you want is called social proof. This is a bias we have to want to do what others are doing simply because they are doing it. I had to be aware of this persuasive tactic that was pushing me in the too-much-tech direction of the masses—and push back.

Most of us don't like being told what to do, and I'm no exception. Silicon Valley may make a great sales pitch, but I've reactivated my very human, very rebellious you-don't-know-what's-best-for-me streak. I pause. I think. I ask questions. I do my research. I consider what's important to me. Then, once I understand what is being offered to me, either to purchase or participate in, I weigh the pros and cons of yet another distraction to my life. I make a reasoned, well-thought-out decision. It sounds simple. And it is. You just have to do it, to take the time to remember your capacity to think for yourself, to get back in touch with your humanity and all of its potential.

Similarly, I needed some TECH TUNNEL VISION.

It sounds like exactly what it is. I run my own race when it comes to my social media and stopped comparing myself (or my approach) to others in that space. I continue to post the things I love, but I do it *my* way. My photos now capture my real-life moments, whether they include working out, my wedding, television appearances, or a peek into a photo shoot with a favorite photographer. I no longer stop my real life to manufacture some moment just so I can post it. There's an organic authenticity to it all.

When everyone is going one way, and you have to—want to—go another, it's not so easy. Groupthink, herd mentality—these are momentum swings and gravitational pulls into which we all get caught.

I have, as I've said, a rebellious streak by nature. Still, I can remember times when I'd get caught up in what everyone else was doing, caught up in the feeling that I had to have something or do something to be part of whatever was happening in the world around me. The social media bubble was one of those times for me.

The key is to take a step back and remember who you are. That requires an acceptance that you may not be as popular for doing things your way. (Or you may be more popular. Who knows?) But regardless, you don't care. Being your own person, finding your own way of doing things in a medium where so many people seem to be following some kind of synchronized pattern, isn't easy. But it's doable. And completely rewarding. When I use Instagram, I use it intentionally, for a purpose. I share a photo or a video because it's memorable or special or useful. I do live videos when I have something to say. When I'm quiet, it's because real life beckons, and that's a beautiful thing. *You'll lose followers! You'll fade out! People won't think you're important anymore!* Blah, blah, blah. I've heard it all from industry professionals, coworkers, you name it. I laugh and go about my day with the knowledge that this life I'm living is supposed to be for me and the people I love, not for everyone else.

When you "just glance" at your social media over and over, you empower this idea that you need to know what everyone else is doing all the time, that you should always be plugged into them, and they to you. You find yourself almost involuntarily comparing yourself, your life, your #MakeupFreeSelfie or #BestVacationEver or #WorkoutWednesday to someone else's, even though you're often well aware that someone's hashtagged moments may not be at all what they seem to be.

The key is to accept that what you see isn't necessarily grounded in reality. Think of the people you know in real life whose images don't at all represent their truth. It took me a little while to fully grasp this whole thing, but once I did, I looked at the medium with very new eyes. It became lighter, more amusing, and I had a newfound appreciation for the people I knew whose online albums were fun depictions of what I knew their lives to be. I finally knew that I was in charge of my own happiness.

I also follow far fewer people on Instagram these days, most of whom I know very well, which makes the whole experience infinitely more fulfilling for me.

Our experience in any medium is what we make of it. With social media, that includes who we follow, what we post, and how much time we spend there.

## CHAPTER 7

HELPING HANDS HOLD NO PHONES

What can we do to be better people in
these fragmented times?

I CAN'T BE THE FIRST PERSON TO HAVE HAD THIS THOUGHT,
this burning question: Why are our devices called "cell phones"?
Last I checked, cells were strictly the domain of living, breathing organisms. What mastermind of the underworld decided that
"cell" should be what we call these gadgets? I can hear the conversation now:

**EVIL VILLAIN WHO WANTS TO TAKE OVER THE EARTH & VAN-
QUISH HUMANITY:** "I have an idea."
EVIL VILLAIN'S SIDEKICK: "What is it?"
VILLAIN: "Let's make a little machine to replace the telephone."
SIDEKICK: "Can we do that?"
VILLAIN: "Of course. We have big brains, we can do anything,
even make other people forget they have brains at all."
SIDEKICK: "Sinful, I like it."

VILLAIN: "And this little machine will fit in people's pockets, in their hands. It will become so much a part of their lives that it will be like a security blanket, always attached to them."

SIDEKICK: "Yes, yes, and perhaps if we make it super-duper easy to use, they'll even give it to toddlers and babies to keep them quiet."

VILLAIN: "Like a pacifier."

SIDEKICK: "Yes. Perfect. The Pacifier of the People, opiate for the masses."

VILLAIN: "That's right. In fact, we'll make this gadget so user-friendly that the humans will want to, and then have to, use this little device every day, all the time, for everything."

SIDEKICK: "Every day? I can't imagine. What will we call this thing?"

VILLAIN: "That is where my true genius comes in. We will call it a 'cell phone.'"

SIDEKICK: "Why?"

VILLAIN: "Because it will be so relied upon, so revered, that it will function like a physical part of the people, like a biological cell. People will feel less than whole without it. One day, the phones may even become the people, and the humans, well, we'll see about them."

CUE EVIL LAUGHTER: "Ha-ha-ha-ha-ha."

[FADE TO BLACK.]

Okay, so maybe that's not how it happened, but you have to admit, there is something sinister about the fact that we call these manufactured gizmos "cells."

Apparently, the real genesis of the name was a tad less exciting than my conjecture. According to a September 2011 article on Gizmodo.com, "The word cellular, as it describes phone technology, was used by engineers Douglas H. Ring and W. Rae Young at Bell Labs. They diagrammed a network of wireless towers into what they called a cellular layout. Cellular was the chosen term because each tower and its coverage map looked like a biological cell . . . Eventually . . . mobile phone and cellular phone . . . became synonymous, especially here in the US. But some people disagree with that usage. They consider the term 'cellular phone' to be a misnomer because the phone is not cellular, the network is. The phone is a mobile phone and it operates on a cellular network."

The phone isn't cellular, the network is. That vast web of connectivity that is taking over our lives is the thing that resembles a structure that is cellular, cell-like, lifelike, living.

I don't feel any better.

MANY OF US BELIEVE THAT WE HAVE A SPIRITUAL CONNECTION to something, somewhere, somehow, that goes beyond the matter of molecules and cells, and certainly beyond the bits and bytes. We are grateful for the rights and privileges that define what it means to be human and would fight to defend them. Yet, here we are, in a battle for our brains, and few of us are fighting. Most aren't even questioning.

When the team at Motorola decided to create a convenient way to talk to others, they had in mind a "mobile phone," as in something that could travel. But could they have imagined that they would create an object that would eventually be everywhere, that people would feel hopelessly lost without, that would be called a "cell" as if it were one of the very components that make up an organism, or that one day could have a mind of its own?

Many days, I battle my computer. I swear, it's like it knows I'm writing a book exposing the perils of tech overload, and it's angry. Peeved. Spiteful. It's determined to thwart me, to frustrate me, to drive me up a wall, out the window, and into the East River. My smartphone? That thing is way smarter than it ever needed to be.

Take one day in 2017, for example, when I woke up to find that my iPhone had somehow updated its operating system, which I didn't want it to do. Did I press something by accident? Did it do it on its own? Who knows. Hours later, I'm watching a movie and hear it buzzing on the coffee table. I go to turn it off and find a new notification that reads: "You have a new memory. Portraits of . . . 2016." There's a little photo of me standing next to someone I don't talk to anymore. WHAT? Was that photo even in my phone? I thought I had deleted everything I didn't want anymore. Had I missed that one? Even so, why was my iPhone pulling up old photos on its own, unsolicited? What was this feature? I didn't opt in, and I couldn't figure out how to shut it off. I went back into my phone, found the old image that I had missed in my deletion sweep, and got rid of it. Regardless, why was my phone acting like a memory prod when I hadn't asked it to?

I imagined all the potential hazards to this monstrosity of a feature.

Think about your phone. Picture a date night. You're with a girl you've been seeing for the past two months. You like her. You're not dating other people, and you have a great feeling about this one, even though you haven't officially said, "You're my girlfriend." You are cooking her dinner. You tell her to grab your phone to check the recipe. She does, and, oh boy, yikes, what the heck, there you are in a "memory" alert and photo with your ex-girlfriend from three years ago.

Mood shift.

Questions arise.

The night isn't quite the same.

And for what? Because Apple decided, without your consent, that your present needed a reminder of the past?

Exes aside, what if a memory alert and picture of a beloved but departed relative you're still grieving over or a pet you miss terribly suddenly appeared on your phone at a time or place when you weren't ready to process those emotions?

Whether it's a photo you never want to see again and thought you deleted, or a photo you treasure but plan to revisit when and where you deem appropriate—why is Apple suddenly calling the shots and determining what you need to be reminded of and when? Frankly, why is it any of their damn business? It's almost like they're saying, "Gee, what can we get away with now?"

Too often, we let them get away with everything. What was next?

# #

ONE MORNING, I WOKE UP TO WHAT I THOUGHT WOULD BE a regular day. I was getting ready for work when all of a sudden, I heard Jeremy shout "Holy shit!" from the kitchen.

Concerned, I peeked out of the bathroom while getting dressed. "What's going on?"

His face was some combination of shock and horror. "Look at that!" he said, waving his hand up and down, pointing out past the window. I threw on a shirt, ran into the living room, and looked past where he was waving. Something big and gray was hovering about, just outside our window. I gave my brain a second to make sense of what looked like a humongous mechanical bug.

"What the . . . ? Tell me that's not what I think it is."

Yep. There it was, right outside my window, with no warning whatsoever—a drone.

Jeremy was now yelling. "And that's a camera! What the hell is happening?"

My heart leapt. "You've got to be kidding me." The camera was peering right into our apartment, scanning back and forth. Then it went down the side of the building, pausing as it peered into other windows.

We tried to quickly snap out of our shock and stop staring at the intruder that was staring at us. Jeremy took multiple photos of the drone, as well as of the two men down on the street who were controlling it, then pulled down our window shades.

Then he called the police.

"Hello, NYPD, how can I help you?"

"Hi. There's a huge mechanical drone outside my apartment window."

"A drone?"

"Yeah, hovering outside my window, staring in, with a camera."

"I see."

"They can see, too," Jeremy countered.

The police arrived. The building manager wasn't in yet, as it was before 8:00 a.m. We had no choice but to leave for work and let the police do their job. Jeremy called a short time after for an update.

"Sir," the officer said, "they have a permit. It's all clear."

"Clear? You're telling me that if they have a permit, they can fly a drone outside our window with no notice and look into our apartment with a camera?"

"You have to talk to the building management, sir."

"Is there no privacy anymore?"

"I understand your frustration, but this isn't a police issue if they have a permit."

"Okay. Thank you."

As it turned out, the building had arranged for some kind of inspection. They claimed to have had no idea that the contractors were arriving that day to do it. They apologized for not notifying us. I ranted, objected, cited civil liberties, did and said everything you'd expect a freedom-loving libertarian-conservative to do. But the case was closed.

Welcome to the reality of life today, where someone with a permit can fly a drone right on up to your living room window. If your building forgets to tell you, that somehow becomes your problem. What is recorded, if and how it's stored, what you might be doing or wearing (or not wearing) when that giant mechanical bug sweeps on up to say hi—none of that seems to matter.

"It's legal."

"They have a permit."

"Oops, sorry, we forgot to warn you."

The whole thing was mind-blowing in a terrifying, what-the-hell-is-happening-to-us? kind of way, one that had me wondering what happens if and when these cells start thinking for themselves, acting on their own intentions and not ours. Then who's really the boss? I started to wonder what would happen if the tech we created got away from us in a scary way. What if humans lost total control?

Apparently, I wasn't the only one wondering.

On the website of the *Independent*, I came across a November 7, 2017, story titled "'Killer Robots' That Can Decide Whether People Live or Die Must Be Banned, Warn Hundreds of Experts." The article begins: "Hundreds of artificial intelligence experts have urged the Canadian and Australian governments to ban 'killer robots.' They say that . . . the development of autonomous weapons will result in machines, rather than people, deciding who lives and who dies. Such systems, including drones, military robots and unmanned vehicles, should be treated in the same way as chemical weapons, biological weapons and nuclear weapons, they say."

The implication is that there could be hundreds, thousands—heck, hundreds of thousands—of these big and little machines placed all over the world. These machines could be controlled by one single, solitary person in a room with his hands on the keyboard, his headset on, his eyes glazed over. Tell me, who is that person? What happens if that person goes crazy, gets power-hungry, or presses the wrong button? What happens if the machines stop responding to human commands because of a glitch, a hack, a who-knows-what?

Toby Walsh, a professor of artificial intelligence at the University of New South Wales, Sydney, was quoted in the *Independent* story as saying: "Delegating life-or-death decisions to machines crosses a fundamental moral line—no matter which side builds or uses them. Playing Russian roulette with the lives of others can never be justified merely on the basis of efficacy. This is not only a fundamental issue of human rights. The decision whether to ban or engage autonomous weapons goes to the core of our humanity."

YES. To the core of our humanity. ON A CELLULAR LEVEL.

All of this might creep up on us sooner rather than later. As Professor Walsh went on to say, "It's not the Terminator that experts in AI and robotics like myself are worried about but much simpler technologies currently under development, and only a few years away from deployment." In other words, these "killer robots" may not show up as scary-faced giant robots with big arms and legs, like the ones we see in the movies. They may catch us by surprise in small, seemingly harmless forms with immense power for destruction.

As an advocate for limited government and greater individual freedom and personal responsibility, I have to wonder what I would do in a world where a potentially government-issued robot could decide my fate or anyone else's. Dressing up as Sarah Connor for Halloween was cool and all, but even that snazzy costume, coupled with my four-day-a-week intense fitness ritual, isn't going to cut it against a militarized, computerized, mechanized heaven-knows-what.

Humanity, and our power to decide what is moral, healthy, and appropriate, is under attack from the developers of artificial intelligence. We're now at a time where people are concerned about autonomous robots crossing moral boundaries in matters of life and death.

Life and death. Sit with that for a minute.

IN THESE DAYS OF TECH OVERLOAD, I OFTEN FIND MYSELF thinking about what it means to be alive—to feel and experience life—as a human.

To me, it means being engaged and aware. It means making eye contact, having backbone-building moments of courage as you say what needs to be said to someone's face, sharing exchanges of facial expressions that reveal your humanity, contributing your thoughtful presence to a moment. I had the advantage of growing up a bit before the technology wave. My relationships were far less fraught with outside distraction. I had the chance to figure out how to approach life's challenges with no tech-enabled easy ways out. I learned how to communicate face-to-face with consequences—good, bad, and character-building. I was experiencing life inside of life itself, with other people doing the same, sensing their feelings, hearing their words, absorbing their energy, contemplating my own. That felt human. That felt alive.

Although, some say that you never feel more alive than you do when you're close to death. When I thought more about my life, it occurred to me that I felt most human, and felt the humanity of others stronger than ever before, on a dark, horrible day, almost two decades ago. But first, a little about how I got there.

IN COLLEGE, AS I MENTIONED, I THOUGHT I WANTED TO BE-come a professor. I applied for and got into the PhD fellowship program in Spanish literature at Columbia University in Manhattan. It seemed like a good idea at the time. A free education with a stipend—standard practice for doctoral programs—at an Ivy League university, to study a subject in which I was interested. I was drawn to the chance to spend my days among old books in the library and my nights reading in coffee shops, absorbing life and literature in a second language. It seemed like a no-brainer.

Turns out, it wasn't the best idea for me. My undergraduate experience had been really good. Maybe too good. From day one, Wagner College had somehow felt like home. The campus was generally quiet, almost pastoral at times. Students often greeted each other by name as they crisscrossed paths. I knew my professors well, the buildings were few, the classrooms were quaint and inviting, and there was just something about the whole place that made me feel like I could find myself there. I met two of my three lifelong best friends there. When I think of my years on that campus, I feel warmth.

Two months into my program at Columbia, the energy just didn't feel right for me. The city and the campus were distraction-heavy zones. Nameless faces passed each other hurriedly without a word. There was a distance to it all, a formality, a sense that I would emerge from those years smarter and more together but not bonded to the people, the place, or myself in the way I had hoped for. Everything about the school was first-rate, top-tier. Yet, unlike at Wagner, I was feeling more lost than found. Whatever it was that I had been looking for wasn't there.

I also knew that the five-year PhD program wasn't for me. The course material just wasn't right. I found myself doodling creative writing ideas, poems, and the beginnings of short stories in my Spanish literature classes. The idea of being "Dr. Bila," a title that still sounds pretty cool, slowly retreated. If I was going to be a doctor of anything, I wanted it to be of something I loved. I decided to finish out the first year, which would complete my master's, then head out into the world to see what was waiting to be discovered.

In the summer of 2001, upon packing up to exit the Columbia program, a friend of my mother's told me about an opening at a large insurance company in downtown Manhattan where she was working. Instinctively, I felt that the corporate world wasn't for me, but I figured it couldn't hurt to try. I inched my foot in the door, got the interview, and soon accepted their offer to fill the position of marketing associate. I had no idea what that meant.

The job was fine. I was a hard worker. My boss and his assistant were nice. I showed up every day and made it a practice to do the best I could at every task requested of me. I spent most of my lunch breaks daydreaming about what I might do next, but the job was stable and paid well, and that counted for something. I told myself that I'd stick it out for at least a year to save some cash. Just a couple of months into the job, on that fateful Tuesday morning—one of the worst days in our country's history—my perspective and priorities abruptly changed.

September 11, 2001. The day was bright, clear. Our offices on the thirty-second floor of One Liberty Plaza sat a stone's throw from the Twin Towers. As usual, I got to work a little earlier than others. I stepped into my cubicle and the landline was ringing. My boss and his assistant didn't arrive until later, so I typically had some time in the early morning to casually eat some breakfast and prep for the day. I threw my bag on a nearby chair, oblivious to anything else on it, grabbed the phone, and gave my usual business greeting. Only, it was my best friend calling. He was supposed to meet me for an early lunch that day but had missed his bus and was going in late, so he was calling to cancel. He started telling me a story about his dad, which was hilarious, so I settled in to listen for a little while, arranging my desk, getting out my breakfast cereal, until BOOM—

He heard it, too. "What was that?"

"I don't know."

"That was loud."

"I know, but I don't see anything from here."

I looked around the office. Most of the staff was scurrying to grab their things, panicked.

"I've got to go."

"Wait, AJ, where are you—"

"I'll call you in a few. Something weird is going on."

I hung up the phone, picked up my bag, saw a manila envelope that had been underneath it, threw it in my bag, and started out. The stairwell was already crowded. I knew it was always best to take the stairs over the elevator in emergency situations, but there were too many people in those stairways and I felt an urgency to get downstairs fast. I decided to take my chances in the elevator. I pushed "1" and stared up at the numbers as they went down. It zoomed to the first floor without stopping. Once I was in the lobby, my cell phone rang. I pulled it out of my bag. It was my dad calling from his office in New Jersey.

"You okay?" he asked.

"I'm in the lobby. I heard a big boom sound. What's happening?"

"It looks like a plane accidentally hit one of the Twin Towers."

"Accidentally? That doesn't sound right, Dad."

"You're okay. It's going to be okay. Stay calm and be careful. If anything else happens, get out of there quickly and carefully, and call me."

"Okay. Love you."

"Love you, too." I hung up the phone. Security was telling us to stay in the lobby and not to go outside. I began pacing. I looked around at dozens upon dozens of terrified faces. I started to feel claustrophobic and had to get out of the building. Outside in the plaza, I looked up at the Twin Towers. One of them had smoke billowing out of the top floors. I couldn't believe what I was seeing. I watched in horror and disbelief as another airplane flew right into the other tower, crashing through its walls, bursting into flames.

I was startled, fixated, confused. Then my instinct to flee kicked in. I ran. As I ran, I tried to call my mother on my cell phone, but couldn't get through. Then my dad. Nothing. Then my best friend. The lines were dead. I kept running and joined the thousands of people on vehicle-less New York City streets, heading east away from the wreckage. A girl to my left tripped and latched onto me. We locked arms and ran together. She was crying; I was in total shock. Everyone was moving quickly, destinations unknown. The girl and I jumped into a taxi, hoping to take the highway north, but the taxi barely moved in five minutes' time, so we paid the driver, got out, and continued to walk north. My cell phone rang. Now the line was working. At least for this call. It was a friend who worked at an airport in New Jersey. He said that my parents had been trying to call but couldn't get through and that he would let them know I was okay. I still couldn't dial out to anyone.

Panicked conversations could be overheard, intertwined with the sounds of police sirens, people crying, the staccato click of shoes on pavement, and the brush of bags as people bumped into each other, apologized hurriedly, then continued on. I had no idea where to go, how to get out of Manhattan. I was starting to feel terror rise in my blood, the fear that there would be another weapon hitting us soon, landing who knows where. I was worried that my parents and best friend were worried. A stranger overheard me talking to two other strangers and offered to let me use her landline in the East Village to call my parents. I thanked her, and the four of us headed north together toward her apartment. We talked about what was unfolding before our eyes, cried, shared fears, and strategized worst-case scenarios. I remember stopping because I could see what looked like little white towels waving outside the windows of the upper floors of one of the towers. I knew that they represented people calling for help. My heart sank. I didn't know what to do. We all stared up in silent grief.

The first tower crashed in on itself and crumbled to the ground. We froze in horror, grabbed onto each other, then ran together, unsure of what we were running from or toward, our only solace the fact that we were in it all together.

A couple of miles later, I entered the stranger's apartment, the one who had offered me her phone. I made two calls, letting my mom, dad, and best friend know that I was still okay. Confused, startled, sad, scared, and grateful, I thanked my new friend for access to her phone and her apartment. We sat for a bit, the four of us, in the solidarity of those moments. We didn't say much, mostly staring at the television as news reports rolled in of the terrorism and evil that had changed our city forever. We would occasionally look at each other, at a loss for words but with a shared expression, knowing that somehow having each other there meant everything. Eventually, we said our good-byes. I headed out with one of the girls, and we walked over the Manhattan Bridge. On the Brooklyn side, I got picked up by a family friend, the girl by her boyfriend.

I spent the ride home unable to say much, thinking of the strangers who had saved me in many ways that day, including the comfort of their mere presence beside me. None of us felt alone. These many years later, I wonder how many horrific, tasteless selfies would exist with the tragic scenery as a backdrop, if selfies or smartphones had been a thing back then. I wonder how many endless tweets would go out with guesses as to what was going on, misinformation spreading like wildfire through the virtual ether. It seems too often now, when there is a tragedy, that the phones come out, texts are sent, photos are taken, posted, commented on, and the event is endlessly debated on social media until it is no longer a multidimensional experience that involves real people, real awfulness, and tragic loss, but rather a two-dimensional political hot potato adjusted to support either side of that day's polarizing partisan debate. One thing I will never forget about that painful day was that there was a deep sense of human connectedness, which was pivotal to our survival and to our ability to emotionally withstand the tragedy that was unfolding.

On the night of September 11, 2001, when I finally got home, I found out that my boss had been frantic all day. He had thought I'd been in the towers. The manila envelope that had been sitting on my work chair that morning, the one I had tossed my bag over to pick up the ringing phone in my cubicle—the one I stuck in my bag before running to the elevator when I heard that big boom—had a note on top from my boss, a note I had completely missed.

It read, in his cursive handwriting: "JB, please deliver this morning to our meeting in the North Tower, 102nd Floor."

Looking back today, I think about what would've happened had my boss been able to send an email or message directly to my smartphone. I would not have missed the message. I would have delivered the package to the 102nd floor of the North Tower that morning. I'm not trying to make a critical, sweeping statement on technology as it relates to life or death here in any way, as I'm sure many in dire circumstances have been helped or saved by technology. But when it comes to my life and how that day unfolded for me, the memo that I missed is something I can't help but think about.

Like everyone in our country, I was horribly sad after, and am still painfully sad about, that horrific day. Like everyone in the city, I was shaken. Thank goodness there are no live photos of burning, falling towers in anyone's iPhone that can pop up as "Memories—2001."

IN ADDITION TO BEING ETERNALLY GRATEFUL THAT I SOME-how survived that day, I was also reminded that life is short and fragile. From then on, I wanted to surround myself with things and people that had meaning and enriched my life, knowing full well that tragic curveballs could hit anyone at any moment. Suddenly the job that was just "fine" but not fulfilling wasn't an option. I wanted to turn daydreams into reality. I quit the insurance company soon after and started waitressing at a Manhattan restaurant while I explored some acting classes. I remember waking up that first day, heading for the American Academy of Dramatic Arts, feeling like I was on my way to finding myself again. I loved the voice, move-ment, and scene study classes. I remembered how to breathe deeply, how to be in touch with my body and mind, and how much I loved getting lost in a character and bringing that character to life. I had one teacher, Vicki Hirsch, who said that sometimes, while playing someone else, you discover yourself. She was right. She also told me that she had a good feeling I was going to be on her TV one day, which I shrugged off with disbelief at the time.

I booked a few jobs, went to Los Angeles, made some great contacts, and even thought about moving there. But there was some-thing about the TV business that gave me pause. At the time, for me, there was a superficiality to it, a darkness, that I didn't like. I stepped away from it to teach and administrate in a New York City private school—the experience I mentioned earlier—as well as to do some writing.

My return to writing felt like being reunited with an old friend I missed dearly. We fell right back into our comfort zone, my pen and I, and I suddenly began to feel grounded in myself, fulfilled, whole. When a different facet of the TV business came beckoning a few years later, after I had written some columns that got noticed, I looked at the whole industry with a different eye. The business was still dark, but *I* felt light, as if I could walk through it all and not let it get to me. It led to some unpaid television appearances, then a paid TV contributor position, a cohosting gig, and other wonderful opportunities. The business hadn't changed, but I sure had. That made all the difference.

#

EVENTUALLY, I DISCOVERED THE ANSWER TO MY QUESTION: *What can we do to be better people in these fragmented times?*

It was this: *Life is short. Be a human being.*

At the very least, in these difficult days of too much tech and too little time, we must REFLECT AND CONNECT. The fact that I survived on September 11, 2001, was not and is not lost on me. But the one thing that really bothers me today, upon revisiting it all in the writing of this book, is that I didn't carry that experience with me enough—that the solidarity of those lifesaving moments with strangers and the importance of our eye contact and our humanity didn't stick with me enough—when I was faced with the growing tech boom that was overtaking my life. I should've been more aware of what could be lost, should've been more in touch with the fragility of life and the need to savor real-life moments, should've been less sucked into some virtual, plugged-in nonsense and more in touch with the people and things right before my eyes.

But it's never too late. I vowed to hold tight to the lessons of that tragic day. And to remember them vividly when faced with anything in life that could threaten my humanity.

## CHAPTER 8

FLESH & BONES & BATTERIES

Why do so many of us seek immediate gratification?

PICTURE THIS:

It's a cool, autumn Saturday night in October. The guy you've been dating since Labor Day has invited you over to his townhouse for dinner with some of his friends. You like this guy. You've had fun together. He's smart, ambitious, and funny. You look forward to meeting his friends.

You ring the doorbell and he kisses you hello, takes your coat, and quickly introduces you to the couple who arrived just before. Trying not to be too obvious, you look around. You're impressed. His home is neat and nicely decorated. The books in his bookcase are even interesting.

Everyone has arrived. Dinner is served. His friends seem nice and easygoing. The food is delicious, you're complimented on the salad you brought, and the conversation is entertaining.

Dessert is brought out. You lean over to the object of your

increasing affection and ask to use his bathroom. He looks a little embarrassed when he says that the one on the first floor is broken, then directs you to his second floor. You go up the stairs, turn down a long hallway, and pass a closet, the door of which is slightly ajar. You can see that it's filled with sports equipment. Next to it, there's a room that looks like an office, with a metal desk, a swivel chair, a framed Picasso print you recognize from a college art history class, and large windows that promise an incredible view of the city. You can't help yourself. You step inside for a closer look. It's gorgeous. The city looks like a postcard. The view was worth that small intrusion on his privacy. You turn back to leave, and WHOA!

What the—

She's staring at you.

And she's naked.

Holy synthetics and silicon. The guy you've been kissing has also been kissing a life-sized sex doll.

This isn't my story, and it may not be yours. But it sure as hell could be.

# #

THEY READ LIKE A HORROR FILM, BUT THESE HEADLINES are real.

"A Sex Robot Appears on a British Morning TV Show"

(Mashable.com, September 13, 2017)

"World's Only 'Talking' Sex Doll Has 18 Personalities, Answers Your Questions and Even Remembers Your Favourite Meal"

(DailyMail.com, April 3, 2017)

"Hackers Could Program Sex Robots to Kill"

(*New York Post*, September 11, 2017)

Sex robots. This is where we are.

Tech in the bedroom in the form of inanimate toys and small gadgets is nothing new. But now technology can take over our intimacies in a whole new way.

Welcome to the age of life-sized, artificially intelligent robots as sex partners, designed to the ideal physical specifications of the buyer. Robot companions that can converse with and fawn over their buyers on a constant basis with no difficult questions about finances or work, no biting comments about the bad-influence best friend, no conflicts with the in-laws.

This is big trouble we're in.

In the early stages of the Internet, pornography exploded. Although, according to a Reuters article published on September 16, 2008, 2008 was the year that "social networking sites [became] the hottest attraction on the Internet, dethroning pornography," pornography is still more prevalent, abundant, and easily accessible than it ever was in any society. Add to that the idea that robots can replace humans in even the most human of circumstances, and it makes me wonder if too many of us have altogether stopped thinking about what it means to be actual, real-life people living real-life lives.

Who would have thought, at the start of the twenty-first century, when we shot forward into the age of Artificial Intelligence, that the person you might be dating could be having sex with a life-sized, computer-programmed robot on the side?

In a December 19, 2016, article on the *Express* online, the headline read: "Sex Robot Shocker: Almost HALF of All Men Will Use Erotic Robot Playthings, Says Survey; FORGET the Notion That Only Lonely Men Are Likely to Buy a Sex Robot."

Madness.

What is all of this doing to us, to society, to our connection to others, to our connection to ourselves? There was a time when you met someone, asked him or her out *in person*, went to dinner and a movie without any nearby buzzing device, got interested in each other, time passed, and you were off to explore tactile, engaged, physical and emotional intimacy with another being, a *human* being.

Now we have machines not only constantly interrupting people, but in some cases actually replacing people. At some point, will the question be: Why bother to engage with real people at all?

We're already glued to our devices anyway. We miss half of what people around us are saying while compulsively checking our emails, social media mentions, and Instagram likes. Anyone, of any age or temperament, can have easy and immediate access to free pornography in the form of any and every kind of video. Graphic images and videos objectifying women and men are everywhere. Idealized, fashion-driven body types are the default we see and expect, even from regular people on social media who are photoshopping images and staging poses. Immediate gratification trumps pursuit and process. We don't give the imperfect guy we liked and had a few dates with a chance because Tinder told us there are fifty other cute ones in a ten-block radius who might be better. If those are imperfect, there will be fifty others. And they all look so good in their Facetune-modified selfies.

Why the heck would anyone bother to take the time to get to know another person, cultivate the patience to work through differences, invest in the promise of fidelity and the commitment to grow together step by step, when, in a fraction of that time, someone can watch free porn, sample neighborhood singles, and keep complicated things like intimacy and vulnerability tightly locked away? I mean, all that effort and commitment would probably get in the way of text flirting and communicating with a bunch of strangers on a bunch of apps anyway, right?

Sigh.

So where does that leave us, as a planet, as a society, as individuals?

Skills like reading social cues, engaging in quality conversations, and interacting in person are being lost. Intimacy, trust, and investment in people and moments are getting drowned out by distraction and instant gratification. All the Instagram likes in the world don't leave you feeling fulfilled. Neither does a rotating door of hookups that are invested in you as little as you're invested in them.

Tech is programming us to be less engaged in real life. Less attached to the people and things in our homes, schools, and families. More fixated on some virtual world full of "connections" that aren't connected at all.

There's just one big problem. I'm still a person. You're still a person. We, the people, are still *people* who crave the intimacy and genuine connectedness that are becoming extinct. The real-life, challenging, vulnerable, committed, present, complicated, full-of-effort moments that seem so hard in the real world and are so easily avoidable with tech submersion are the ones that make life worth living.

#

THE MASHABLE ARTICLE REFERENCED AT THE START OF this chapter, "A Sex Robot Appears on a British Morning TV Show," featured a clip of the actual British talk show segment where hosts Philip Schofield and Holly Willoughby of *This Morning* interviewed a sex doll's inventor. The inventor sat next to his creation, sex doll "Samantha," on the couch, her hand on his knee and his hand intermittently on hers. "Samantha" lives with him, his wife, and his two children.

Willoughby's questioning of "Why, why, why? Why is this necessary?" echoed my own thoughts.

"Samantha" was petite, with long brown hair and delicate, hourglass proportions. She wore a red dress and had facial color tones that resembled makeup. The inventor often touched her gently, as if to show affection toward his precious piece of wires, plastic, and silicon. His wife stood off oddly and passively in another area, arms crossed, her expression a seemingly nervous smile as she defended her husband's mechanical mistress. The fact that the doll was beside him on the couch while his wife was off in the wings struck me immediately. I wondered if his wife's positioning in that segment symbolically represented her second-tier status in the marriage. Next to the doll on the couch was a psychologist and therapist, Emma Kenny, who was as visibly uncomfortable as the hosts while the inventor talked about how lifelike his doll was. He touched her skin and described the different textures that made it feel real. Schofield said he touched the doll, too, before the show, and it chilled him: "She's like a corpse." The inventor explained that she could one day feel warm, like a real human body, if they could increase the power of her battery pack.

The whole thing freaked me out like you can't imagine. Or maybe you can. It was like watching a grown man bring his favorite oversized stuffed animal to work, only somehow he had forgotten that it was just a teddy bear and he was treating it like a person. This was even worse. Because this was a sex robot made to look like a real woman, and he was touching it and talking to it and engaging in heaven knows what with it while his actual wife was heaven knows where. I couldn't compute it all.

Schofield put it well: "It's a bit like making love to a car GPS, isn't it?"

As if it couldn't get any worse, the inventor then went on to explain how the doll can go into "family mode" so that if she meets the family, she can talk about animals, philosophy, or science, and even houses a thousand jokes. "There's a lot to Samantha, she's advanced," he declared. He then shared that his three-year-old and five-year-old ask for "Samantha" when she's not around.

The hosts and the psychologist seemed as horrified as I was.

"But at some point, [the kids] are going to go, 'I'm now old enough to realize that Daddy has sex with Samantha, and Samantha is not Mummy.' Is that not a bit strange?" Willoughby asked.

Unfazed, the inventor answered: "No, I think the world's changing."

Changing into what, exactly?

The psychic, I mean psychologist, read my mind: "But it doesn't mean it's changing for the better, does it?" she said. "And I mean I think, with AI particularly, one of the things that we're realizing in psychology is that sometimes just because we can do something doesn't mean that we should do it. And I think sex dolls are a perfect example . . . We're objectifying women, but worse than that, we're commercializing them, becoming consumers of women's bodies . . ."

Then the therapist seemed to lose any tolerance she had managed to muster. Her voice escalated. "This is not real, she is not called Samantha, she is a piece of whatever you've made . . ." she charged emphatically, until Willoughby cut her off. The interruption seemed the likely result of a producer's prompt to offer some balance to the segment, because Willoughby then said something through which it seemed she was biting her tongue: "But are they doing any harm?"

Yes. They are. And then some.

Of all the things to touch with a heavier and heavier hand of technology, why sex? The height of intimacy, many would argue, the point of physical connection with a HUMAN BEING that can, under the right circumstances, actually produce another human being. I can't believe I have to even type this, but ROBOTS AREN'T REAL. If we define being real as the ability to copy texture and temperature and forget about the intangible, emotional, sensual waves of thought and feeling between us, then, really—what are we doing these days? What kind of lives are we living? At what point do we, with our purchasing power (hello, free market) and our desire to preserve the value of what's real, of what it means to be intimate, to interact, to connect—at what point do we stand up and say no?

#

THINK ABOUT WHAT TECH HAS DONE TO DATING. OR, BETTER said, what we have *allowed* tech to do to dating. There are so many dating apps and dating options, such a flood of people right at your fingertips every day, that many are hesitant to zero in on a special person and settle down. People constantly ask themselves, *Who else might be out there?* We flip through romances the way we flip through television channels, unwilling to commit to one show, one person, for fear of what we might be missing.

Apps like Tinder and Tinder Plus, Bumble, The Grade, Hinge, and Happn promote the possibility of meeting as many people as fast as possible, wherever you are, in pursuit of the optimal casual hookup. No strings attached. A quick fix. Rotating doors of people who barely scratch the surface of intimacy. If that's what you want, then it's your right to pursue it. But let's be honest about the fact that these apps encourage that instant-gratification path, invite you in, and hope you stick around—with the app, not the person.

There is even an app that, as BuzzFeed News put it in a piece on June 20, 2017, "lets you find people on Tinder who look like celebrities." This app, Dating.ai, entices users to upload a photo of anyone they want, including old crushes, exes, and famous people, and then scours various dating apps and websites to find prospective dates who look exactly like them. Can you imagine? You put your profile on a dating site, and this other app locates you because you look like some guy's ex. Then he wants to date you because of that.

Come on.

A whole generation of people in their teens, twenties, and even thirties now treats dating, romance, and sex much differently than the generations that came before, in large part due to technology. It's not uncommon these days to hear about high school students who are "dating" and text all through the night, yet rarely talk to each other in person at school. There's also the whole ballooning "hookup" culture—the idea of engaging in physical activity without any romantic feelings or, often, even friendship entanglements—

that's growing because of encouragement and support from social media and dating apps. Sure, the hookup culture has existed for a while, but it wasn't nearly as prevalent or embraced as it is now. When I was in high school, there were some girls and guys I knew who hooked up with random people here and there. But it wasn't the norm, and it wasn't glorified by any means or held up as some form of female or male empowerment as it often is today. Now, thanks to the dozens of social media platforms and apps, and a phone in our hand that can send quick messages through the air at lightning speed to anyone, anywhere, even five feet across the room, a looseness about intimacy has taken over.

Instant gratification.

Floods of options at the click of a button.

Tinder this and Grindr that.

Impersonal texts over in-person eye contact.

"U up?" texts at 3:00 a.m. because it's effortless and you face no fear of in-person rejection. It's been normalized to the point where you not only don't think about how you could be interrupting someone's new relationship, but you also don't consider that you're pretty much engaging in glorified prostitution. Without the cash.

Do young adults still ask each other out on dates? I'm not sure if many even know how. Sure, they can text a big "heyyyy" or whatever, but can they walk across the floor of the gym at a high school dance and say, "Hi. My name is ———. What's yours? Want to dance?" Do they have the character and courage to try it, to face the possibility of rejection without the shield of a faceless, distant, safe-space text? Remember the guts it took to cross that gymnasium floor? The eye contact? The putting-yourself-out-there reward of knowing that even if the object of your affection said no, you were brave enough to try? These are the little things that turned boys and girls into men and women.

What about people in their twenties and thirties? It seems these days even they would rather text a lazy, risk-free "Watcha doin?"

than pick up the phone, dial, take a breath, say hello, and have a conversation with the hope of it leading to . . . who knows?

Remember the old days when people picked up a landline phone to call a friend, went to the store in person to buy food, books, and clothing, visited a travel agent to book a trip, cut out newspaper ads to find an apartment, spoke directly to a bank teller to withdraw or deposit money—went to the movie theater, for crying out loud? I'm not saying people don't still do some of those things, but there are also countless faster, tech-assisted alternatives. Now, life is all about instant, push-this-button, get-it-to-me-here-and-immediately satisfaction. We can do it all with one tiny phone. Soon we'll be able to do it with the blink of an eye. I have no doubt that someone, somewhere, will soon start putting little devices in our heads, like the mind-chip examples I mentioned in earlier chapters that are being tested at Facebook, Neuralink, and Three Square Market.

In his landmark July/August 2008 article in the *Atlantic* titled "Is Google Making Us Stupid? What the Internet Is Doing to Our Brains," writer Nicholas Carr noted that while he appreciated all the information he could get on the Internet, "what the Net seems to be doing is chipping away my capacity for concentration and contemplation. My mind now expects to take in information the way the Net distributes it: in a swiftly moving stream of particles. Once I was a scuba diver in the sea of words. Now I zip along the surface like a guy on a Jet Ski." This was ten years ago.

Superficial consumption of content is constant, but depth of experience is lost.

Easy isn't always a good thing. If we continue to pursue easy, to grab at the easy way out every second we can, we'll wind up losing not just valuable skill sets but valuable lessons, growth, and the internal boost of conquering hard tasks. There's a gift to the struggle.

As Frederick Douglass, the abolitionist and social reformer, put so well in his 1857 "West India Emancipation" speech, "If there is no struggle, there is no progress."

My grandpa used to say something similar, something about how nothing easy to achieve is worth it. As I look back at my life, I can see that he was right. Yet, it took me a while to realize that I was falling into easy as easily as everyone else, especially when it came to using technology to pursue my career.

# #

AFTER SPENDING MY FIRST FEW YEARS IN THE WORKFORCE sampling jobs in marketing, the restaurant business, and finally as a schoolteacher and academic dean, I moved into broadcasting. I started by writing columns and doing lots of unpaid on-air appearances at Fox News, MSNBC, and CNBC. Industry professionals told me that a strong social media presence was "the easiest way to get your name out there and build a brand." I was advised to be very active and very present in that online space. So I listened.

When I first got on Twitter, I was on it way too much. I would read my feed incessantly, followed way too many people and news sites, and would respond to almost everything. I was constantly commenting, and if I wasn't commenting, I was reading someone else's comments or jumping from a Twitter post to a news site to a Twitter post . . . and repeat. Then repeat again. And again. I was under the impression that my success in the field was linked to expanding my voice in the social media arena. I thought, *I'm in a social media age. If I don't do this actively, someone else will, and they will get the job.*

I got a feeling of exhilaration from the immediacy of social media. The more I tweeted out, the more satisfied I was. On Facebook, I approved almost anyone who wanted entry into my world. I didn't even think about filtering. That was on my "Friend" page, a page that, prior to my entry into the TV world, had consisted of only those I knew in real life. Eventually, I also started a "Like" page at the recommendation of a TV colleague. My "Friend" page eventually maxed out with mostly strangers, but the "Like" page had no limits. All comers were welcome.

Soon after, I started an Instagram account. People in the TV business told me I was late to the game. They told me I had to have a *visual* social media presence, too: "It's relatable. It will bond you with the audience in a more personal way." So I took the plunge.

To put it mildly, I was very present on social media.

More than I was in my life.

When I woke up, my daily routine was the same. I checked my phone—which rested on my nightstand—perused my texts, my emails, my Facebook pages, my Twitter, my Instagram. That was before my feet even hit the floor. Then, after I brushed my teeth, off I'd go to the living room, where I'd turn on my computer, head to the kitchen to make a morning shake and breakfast, go right back to my computer to check the emails and news sites I'd just checked on my phone, and then click back to social media.

It was *not* time well spent.

On Facebook, I would find myself landing on my childhood friend's ex-boyfriend's cousin's mother-in-law's rant about the furniture that arrived broken. On my page, I'd see random, often odd comments about me and messages written by people from my past who had no place in my present and whom I didn't feel the need to talk to anymore. They'd be mixed in with some nice comments from TV viewers, and some horrific ones from trolls on their mission to be nasty.

On Instagram, I struggled at first, chronicling everything, taking photos of everyone, sharing quotes of this and that. I was always in "Like" mode, tapping the hearts, reading the comments, curious about what others had liked, following people I didn't even know. There, too, I'd land on strangers' pages that left me wondering why in the heck my friend's boyfriend's sister still had pictures up of the ex she told us had kicked her dog. Two hours could pass, just like that. It was like having an open therapy session with the world. Only I wasn't leaving those pages feeling good.

Then one morning, after a night of reading comments and posts that struck me as particularly mean, I woke up and thought: *Who gives a shit? Why am I doing this to myself?* It was obvious why. I wanted to be in the loop, getting the scoop, building a name for myself, staying in the know so I could move up the ladder, both socially and professionally. I remembered the industry professional who had told me that a strong social media presence was "the easiest way to get your name out there and build a brand." She had been right. Thanks to my hyperactive Twitter presence, my follower counts were continuously rising and I was generating good buzz about my TV appearances and columns. There was just one problem. I had ignored the reality that the way I was managing those accounts and the intensity of my presence there wasn't making me happy. In my effort to take the easy way upward in this industry, I had forgotten one person, and that person was *me*. The instant-gratification route may have had its practical benefits, but it wasn't the most rewarding. So I decided to make some changes.

I still liked and needed many aspects of social media. I liked the ability to share my observations, insights, and news stories immediately and expansively on Twitter. I liked having a digital photo album on Instagram. I liked posting clips of my work or links to articles I had written on Facebook. I didn't want to be left out of professional opportunities that emerged due to a social media presence, so I had to find my own way to deal with these mediums, to navigate my own path, to create some boundaries, and to not be overwhelmed or have social media have a negative impact on my life. I had to figure out how a simple, old-world soul like me could survive and stake my claim in a television and news industry that had social media at its core.

First, I removed a bunch of posts. I didn't want many of those things to be available in a public space. Why was I telling people my frame of mind through a quote just because everyone else was? Why was I tweeting a very personal thought that I should've reserved for an in-person conversation? Why was I concerned about relatability and exposure and what industry people were advising me, and forgetting what was good for me?

Second, I decided I would no longer read comments on my Facebook posts. There was too much nasty littering by trolls there. Monitoring it all would cost me too much time and aggravation in the social media world—time away from real life. I did my best to follow the same policy on Instagram, although I still engage in a few exchanges with close friends. Twitter was a bit different for me, as I sometimes do an interactive Q & A with viewers and readers, and have the willpower to duck in and out quickly when replying to someone here or there, but I decided to scale back my presence there significantly as well.

Initially, it was a struggle. But I had to try. It took a few days, but eventually I started ignoring the comments that followed my images, posts, and most of my tweets. I knew I was lucky to be on television and I was, and still am, humbled by the opportunity. But I was also aware that the exposure on TV made me a no-holds-barred public target for some reckless commentary. Ignoring many of the comments was the only way I could continue to pursue this career.

All of that brainstorming about instant gratification, about what was easy versus what was rewarding, got me thinking about the impact of the tech boom on valuable, dying, once-thriving skill sets.

THE SKILLS THAT PEOPLE USED TO HAVE ARE DISAPPEAR-ing. Of course, as societies move forward, this is often the case. Few of us know how to grow our own food, build our own houses, or make our own clothes. But it seems now that more and more personal skills are disappearing, too.

For twelve years I went to a Catholic school, where penman-ship mattered. Every written word had to be neat and clear. When I recently wrote a birthday card in script to a family friend for her twenty-second birthday, she opened it and looked at me in awe.

"How did you learn to write like that?"

"You mean neatly?"

"Yeah, but also in cursive. I don't even know how to do that anymore. I studied it for a year in lower school and rarely had to use it in school after that." I told her that I had to take penmanship classes and tests for years, and even got report card grades in it. She almost fainted.

But there is something important about writing with a pen on paper. It affects your thinking process. A 2016 article in the *Wall Street Journal* titled "Can Handwriting Make You Smarter?" cited a study conducted at Princeton University and UCLA in which stu-dents who took notes by hand did better on their tests and in their classes than those who took notes on their laptops.

I'm glad I learned the old-fashioned way to write. It was hard and tedious and you had to struggle to get those loops just right, but, looking back, it built so much character because it was difficult. A challenge. And, when you got it right, a victory. Those penmanship classes instilled in me at a young age that if you wanted to commu-nicate something, you had to take the time to do it well. You had to practice every night and perfect the craft, and if you did, then voilà, there it was, your imprint on the world in your very own beautiful script. I still have some of the black-and-white, marble-cover com-position notebooks from grade school where I kept my class notes. I look at them now and almost can't believe the volume of handwrit-ten content, the time I took to diligently fill up those books. In one ninth-grade class, I filled four notebooks, tediously writing at light speed to keep up with my Global Studies teacher, who taught us that if you wanted to rise to the top of the class, your mind and your pen couldn't miss a beat. She would collect those notebooks a few times a year and grade us on our note-taking, the content, neatness, you name it. I look back and wonder if I could even do that now. With all of my keyboard usage, I'm out of penmanship shape. I know I could never write so legibly and quickly. More importantly, would I even have the attention span, the kind of mind-set that could focus exclusively on a task for forty minutes of uninterrupted, rapid note-taking by hand? Or am I so used to so many breaks to check this app or that text or read this twitter post that I've lost my ability to focus?

If there is an easy way out, we tend to take it these days because we are busy and overwhelmed. With technology being what it is and having penetrated so many areas of work and play, we're inundated. But upon thinking about my grade-school composition notebooks and some skills I may have lost, I now take some time to do things the hard way. Like that one time at a roller rink . . .

# #

ONE AUTUMN EVENING, JEREMY AND I VENTURED OVER TO an outdoor roller-skating rink in Brooklyn. It was a great spot, a newly built green space near the Brooklyn Bridge in a park with basketball and handball courts, and even bocce and shuffleboard. I loved it right away.

I had brought my old-school skates, the ones with four wheels. White skates with fluorescent pink wheels, to boot. They're heavy and clunky, but to me they're the real deal. Jeremy rented what any normal person in 2017 would rent: sleek, fast Rollerblades. We sat on a bench outdoors, put on our skates, laced them up tightly, and off we went, into the rink.

More accurately, off *he* went. He skated around looking fabulous, like a pro. I'm a pretty good roller-skater, but he was way faster than me in his blades. I'd glide along, lifting the dense weight of the old-school skates as I rounded the bend of the rink. It brought me back to my childhood, when I'd have roller-rink birthday parties in middle school, '80s or '90s music playing in the background while a giant disco ball twirled and sparkled above.

"Why don't you try the old-school ones?" I asked Jeremy as we stopped to chat.

"Why?"

"I just think you might be missing something."

"Like what?" he asked. "I feel like these are better, no?"

"Just give it a shot," I said.

And he did.

He rented some old-school skates and stood up.

"Wait—what?" he said, as he grabbed the ledge. "Why is this so much harder? It's like you have to balance in a weird way, and they're heavy."

I grabbed his hand and smiled. "Come on, slowpoke, follow me."

After a few times around, we took a break for a snack. Jeremy shook his head. "Those are a lot of work," he said as he laughed. "But I get it. It *is* different. You really have to do the work. You have to pay attention to your feet and your balance way more. To focus more."

Paying attention. Focusing. Lost arts in many, many ways.

Our roller-rink episode got me thinking even more. Improving the quality of things—of anything—may make activities smoother, more efficient. But is that necessarily better? I thought about the ways that technology had made things easier, like GPS taking over for paper road maps, email taking over for handwritten letters, texts taking over for phone calls. Easier, yes. But were those changes necessarily for the better? Is it better to engage in an activity without having to work as hard or put in as much effort and focus? Is it better to not have to use our brains, engage, and pay such close attention? If we don't pay attention, use our brains, put in that focus . . . what might happen to us as humans, as a society?

It can be rewarding to put in work—real work, the kind that taxes your body, mind, and soul. That's why people garden, do crafts and home projects, build things, work out, and run marathons. It's satisfying to put in the time and effort to accomplish a goal. As it stands now, I look at the way we use our technology, the easy access to almost everything, and I can't help but wonder what we have lost, what we stand to lose. I'm scared of what, in ten or fifteen years, today's children will become. Will they understand what it means to build relationships, to physically exert oneself, to mentally engage in problem-solving?

# #

I DISCOVERED THE ANSWER TO MY QUESTION:

Why do so many of us seek immediate gratification?

It's this: *Because it feels good.*

But not for long. We cannot fall for easy-peasy. We must embrace the struggle, which can be surprisingly rewarding.

For me personally, I needed to work on having more PATIENCE because I was still failing the Marshmallow Test more than I cared to admit.

The Marshmallow Test was a famous experiment conducted in the 1970s at Stanford University. The gist of it was that little kids were tested on their ability to delay their gratification—that is, to wait for their reward. In the test, each kid was given a sweet, which was often a marshmallow. They were told they could eat that treat now or . . . they could wait. If they waited, they would get two marshmallows when the researcher came back in the room and be able to eat them both.

Some kids ate the marshmallow right away. Some waited. The kids were then studied over a decade and there seemed to be a correlation between the kids who had the discipline to wait for their reward and better test scores, grades, and physical health.

How many of us would fail the Marshmallow Test if a device, instead of a treat, were sitting right there on the table in front of us?

A buzzing smartphone full of incoming texts, social media notifications, and emails, all begging for our attention? Most of us would fail. We'd be compelled to grab it. Few of us can wait for anything these days. We have no patience, even if it means getting a reward later. We want our emails immediately. Our texts? Some people can hardly stop driving before they read and respond to the most recent, very important message on their phone. Need to make a phone call? If we don't have a phone in the palm of our hand, we panic. The feeling that *we must have it now* is a sure sign that we are being taken over.

Need a pizza, a ride home, someone to fix your windshield? Tap your phone and you'll probably wait less than thirty minutes for any of those. Want to know who won the big game, if it's going to rain, the circumference of the earth? Talk into your phone or some other device in your home, ask the question, and moments later you'll be cheering your team's victory, getting out your umbrella, or sharing with friends that it's 24,902 miles around the equator. Parents have a question for their kid? Why wait until school is over? Send a text. It'll take a minute, and they can answer during study hall. Eager to reconnect with an old friend from long ago? No need to call their childhood home, talk to their parents, or find out what state they are in now and dial information. (Remember all that hassle?) Nah. In three minutes or less, you can probably find them on the Internet or in the pages of social media. Their email may even be listed online. If not, a direct message is one click away. There's a good chance their cell number, like yours, has never changed and is still in your phone. Tap-click-click and you can contact them immediately.

We like to get our needs and emotions satisfied as soon as possible. Instant or near-instant gratification is what's popular these days. But just because we like the immediacy of it all doesn't mean it's good for us.

No barriers between your desire for food and your ability to get it? Then you can eat all day long with little effort on your part, except the chewing.

Facts at your fingertips mean a whole lot of knowledge is available, but how much actual learning is involved? What do we figure out on our own? What skill sets do we build, and what have we lost by taking our journey of discovery out of the process?

Lightning-fast constant communication between parents and kids keeps the tether tight, the leash short, the proverbial umbilical cord uncut. Yet, some kids—many kids—might want (and need) some hours during the day without access to their parents, where they can focus on their teachers and peers, where they can actually grow up and learn to navigate their school challenges and friendship challenges, to problem-solve for themselves with no consistent, one-tap-away parental crutch.

What about that friend from long ago you plan to text out of the blue? You want to reach out, but have you thought it through? Or has the ease of the process, the quick search online and even quicker tap and send, made you act without a care in the world, without contemplating how things left off and what the best way to approach things might be?

Today, with the ease of a tap-click-send world right at our fingertips, restraint becomes a necessity and takes a lot of work. For me, it was a particular struggle. So I set up some small personal rituals for myself in the hope of shifting my patterns.

When I'd get angry or upset at something and get the urge to send a page-long text essay across the ether to someone, detailing everything they did that made me mad . . . I'd stop. And breathe. I'd put the phone down and take a walk around the block, around the office, even just to the kitchen to get some water. Then I'd sit back down, pick up the phone, and type, "Let's talk about this later on the phone or in person." If and when they'd resist by typing back frantic, emoji-packed diatribes, I'd take another lap, breathe some more, then type back the exact same words I'd written before: "Let's talk about this later on the phone or in person." This wasn't easy for me at all, as my personality is prone to wanting to respond instantaneously, with passion, as I had unfortunately done many times before. I'm not wired for patience. I had to battle my tendencies for a little bit, to play tug-of-war against myself. I did it, and it worked. Eventually, people came to understand that I wasn't going to engage in long, deep, serious, argumentative, emotion-packed conversations anymore over text, and so they didn't try. The madness stopped, for the most part.

I now practice the lost art of patience. Even before the age of tech, I always wanted to get things done yesterday. I was the type of kid who would start putting toys together before I had read the instructions. Then I was the high school student who would start exams before the teacher had finished explaining what to do. I can't count how many times I started driving to a destination before I had even figured out what bridge or highway I needed to take.

If I can learn some patience, anyone can. Tom Petty and the Heartbreakers were right: "The waiting is the hardest part." I hate it. I'm not good at it. But I also know that satisfying my need for immediate gratification, my desire to have it all right now, rarely has good consequences.

In addition to no longer having heartfelt or challenging conversations over text—or email, for that matter—I don't text or read emails while driving. Even at stoplights. Ever. I've learned that the universe doesn't internally combust if I wait to check my email until I park the car.

These simple steps got me to start training my tech-flooded, instant-gratification-craving mind to breathe and consider. With that came better conversations and better solutions to problems, more fulfilling relationships, less wasted time, so much less stress, and an appreciation for human capabilities like patience that I had started to forget I had.

## CHAPTER 9

THE POWER OF POSITIVE

How can there be so much hate out
there? Or is there?

ONE DAY, WHILE BROWSING ABOUT IN ONE OF MY FAVORITE
Manhattan used-book stores, taking in the faded bindings, the
soothing scent of worn pages, the other book lovers' slow, out-of-
time perusals down the aisles, I overheard a mom by the register
talking to her son about why it wasn't okay for him to say mean
things to other kids on his Facebook page. I was catapulted back to
the social media age. Bullying. Why does it seem to have gotten so
much worse? That small moment, that observation, compelled me
to think more deeply about what was going on in my own social
media life. Of things that have been said to me online. Of things I
may have said to others online.

From the start of my television career, my libertarian-conservative
positions weren't always loved. I'm a small-government gal, and that
means criticizing Democrats *and* Republicans who spend too much
of our hard-earned cash or restrict our freedoms and civil liberties
when it's convenient for them. I'm a Principle over Party person, an
equal-opportunity critic who isn't interested in preserving a Party

or a politician, but rather in telling the truth as I see it. My truth centers on less government control, more power to the people, respect for our Constitution, and a willingness to take on both sides of the aisle to demand accountability and transparency from politicians. Upon hitting the broadcasting scene, I soon discovered that my approach angered plenty. Many Republicans hated that I was pro–limited government but not a Republican loyalist. I'd call out big-government Republicans with the same vigor that I'd call out big-government Democrats, and some Republicans didn't take that so well. Some liberals couldn't stand me because, well, I wasn't liberal. I was a vocal advocate for freedom in all its forms. Sometimes that meant your freedom to marry whoever you wanted (liberals generally liked that), but other times it meant your freedom to not have health insurance if you didn't want to (they often didn't like that so much). Between my refusal to toe a Party line and my demand for ideological consistency, I've angered both sides pretty regularly. And, thanks to the come-one-come-all access of social media, I've come up against plenty of online hate from viewers.

When I was on *The View*, the other hosts and I certainly had our share of political disagreements. But when the stage manager called it a wrap and the lights and cameras went off, we wiped away the makeup and went back to regular life, talking about an upcoming outside project, our need to get away that weekend, or what we craved for dinner that night. There were always a few dozen people out there on social media who never quite got that. They had the need, the impulse—the burning, raging desire—to put me in my place, to tell it like it is, to make sure I knew that they think I am *less than they are* for offering up an alternative point of view.

So I heard from them. Often.

As an example, this one, sent out about me on Twitter in August of 2017, that a friend brought to my attention with a how-do-you-do-this-every-day? phone call.

The tweet read:

> You r a pitiful racist no good bitch. What did you mean on the view you did not live around black, n u was rise up with white peoples aka the slave masters. Know I know why u r so evil look who you was rise up with, the devil's p.s. Ur not a white girl ur a fucking wowper.

You may take one look at that and tell me I should have laughed it off and dismissed it. I may wholeheartedly agree. But we're all human, and it's not always so easy to do that. I have thick skin, a sense of humor, a supportive network of people around me, and an awareness of the goodwill of my intentions, but still, I'm a person. Sometimes when toxic or threatening comments are called to my attention by well-meaning people, I don't feel like laughing them off. Instead, I'm left thinking about the world we live in that generates them.

The social media departments at most TV talk, commentary, and news shows are a part of the product package. No matter what television show I'm on, I like to set aside some time on Twitter to interact with viewers, to facilitate dialogue, and to generally support the efforts of the producers, executives, and entire production team of that show. Also, I mentioned earlier that I occasionally use Twitter to constructively interact with people engaged in my work, at times offering a Q & A forum where they can ask questions and I can share fun aspects of my life or break down stereotypes of what a libertarian-conservative represents. But when I peruse the mentions and find hateful words directed my way, it does make me question even my limited interaction in that space. For now, I think it's still worthwhile.

I had a similar experience recently using Instagram's live video feature. I was attempting to do a live Q & A, inviting viewers to type questions in real time, but I wound up having to read and skip over too many horrific comments from trolls. So I modified my vision and now use the video feature as an opportunity to speak to viewers without paying much attention to the incoming comments.

When one viewer of a show on which I appeared asked during

a commercial break if I'd ever been trolled, I laughed and replied with "When have I not?" Even the word "troll" is so nauseating—existing for the very act of baiting, provoking, and bullying someone. Regardless, I was pretty sure I could pick any given period of time on social media and give an example of someone saying something nasty to or about me. I went to some of my feeds during June and July of 2017, a portion of the time when I was on *The View*, and found the following. These are verbatim, with no fixes to the spelling or grammar:

> ABC please boot these lame ass chick from The View panel. Get rid of her ass. JEDEDIAH #GurlBye.

> You're awful on the show and I hope you're the next one fired.

> Most of her pictures have her with her mouth wide open . . . scary . . . feel bad for her husband . . . if there is one?

> Please, Jedediarhea is nothing shy of a Fox News bickering bubble-head who will soon be replaced on The View ! I need Imodium just to listen to her screech about Trump every morning. Pathetic.

> can u put a hindge on ur neck n kiss ur own azz. ps quit taking wit ur hands like a crack haid !!

> I actually thought you were a good reporter! Ha! You are just as bad as your stupid friends on the view, the worst show on the planet! Don't bother coming back to Fox news, I will never watch you! You stink!

> #TheView oh please @JedediahBila lives in a loud lie world.

> Jedediah, you are literally the biggest hypocrite on morning TV right now. I have NO clue how you are even on the view. YOUR views are

SMALL, hypocritical, and you contradict yourself every single day! Have you noticed how the audience NEVER applauds ANYTHING you say!? You spread hate and fear and lies day in and day out. You have a total lack of empathy and understanding! I look forward to the day when you are no longer on this show, just like I look forward to the day that the POS fake president is impeached.

Nobody does the model face shots other than you. This show will not make you a star, it only shows how ignorant, bitchy, repeater & stuck up your ass, person yu are. Read The View posts, no one else is pleading for you to be GONE.

I could go on and on.
But then also this:

You do realize she's a real person with feelings, right? What makes you think anyone would want to read that about themselves? So cruel.

But then again, this about Joy Behar's puppy dog, Bernie, who had a guest appearance on the show in July 2017:

Bernie should be a new co-host. He could replace @JedediahBila & do a better job just by showing up. Not to mention, he's a whole lot cuter.

Still want to be on TV?

OUR TECHNOLOGY IS MAKING US LESS OF THE PEOPLE WE need to be: good, decent human beings who think for ourselves and act in ways that are true to who we really are. I can't tell you how many times I've sat in green rooms at cable news networks observing prominent TV personalities preparing for various appearances on different shows. I see these people, who often have very different political opinions, laughing and smiling as they talk about their families and friends, having hearty and thoughtful conversations about politics and culture, and getting along through it all, mutual respect and diversity of thought going hand in hand in these face-to-face settings. Then fast-forward a few hours, and I've watched those same people get nasty and condescending with each other on social media. Sometimes personal attacks are thrown back and forth. It gets ugly, and then uglier.

Two opposite kinds of interactions, one in person and one online, with the same exact people. Why so different? Because in-person exchanges and online exchanges are different worlds.

Face-to-face interactions involve thought, eye contact, an awareness that we're all still people, and an implied accountability for our tone and behavior. Online mediums invite people to lose their humanity, to address people in ways they would never do in person. Online, too few of us take the time to think about our words or remember that another real person is receiving them. Online, our better selves take a backseat to impulsive, often unhelpful and unproductive language. Online, too many of us treat each other like garbage. It makes me think of middle school. The teasing, the jockeying for a spot at the popular table, the insecurity and bullying. That was an awkward, uncomfortable time for almost everyone. But then, weren't we all supposed to learn from it and grow out of it?

Has adulthood not hit the social media scene?

Or is the problem that everyone now feels like it's their job, their duty, to comment on just about everything all the time, as if social media has unleashed a world of social commentators, many of whom would never have the courage to say those words in person? Have Twitter, Facebook, and Instagram created mediums where you're inspired to be active and vocal, alerting everyone to your emotions about everything and everyone all the time? Does the detachment of a computer or phone screen, or even the anonymity of the name or photo from behind which many tweet or comment, allow the worst sides of people's characters to emerge?

Yes, yes, and . . . yes.

I thought being on television would be a lot of fun. And it is. I like many of the people I get to work with when I'm on a show. In my experience, the producers are typically supportive, the writers aware and earnest, the on-air talent insightful and quick-witted. I genuinely appreciate the viewers more than they know. They are the reason I get to do what I do. I feel incredibly lucky to have had, and continue to have, various media platforms on which to share my views. The feedback is sometimes rewarding and encouraging. Like these:

@JedediahBila watched your interview with AbbieLeeMiller and was blown away. Never watched show DanceMoms. You were fantastic. #proudfan.

Loved seeing @JedediahBila (even if only a clip) on @FoxBusiness . . . Miss ya Jed!

I like @JedediahBila. I love that she's not afraid to ask questions and get clarity.

@JedediahBila has such a great laugh #TheView

I said it before and will say it again. You do your homework and that's why I respect your opinion 100% @JedediahBila.

In fact, when I exited *The View* in September of 2017, there was an outpouring of kindness on social media from conservatives and liberals alike, including some who vehemently disagreed with my politics but respected my approach to debate. Some conservatives and libertarians thanked me for often representing their voices at the table. Some liberals thanked me for opening their minds to new ideas and getting them to think about things they hadn't considered before. Others thanked me for being an independent thinker who was willing to criticize corruption and inconsistencies on both sides of the political aisle in the name of principle. All of that made me smile. I had done what I had set out to do—to shake things up, to question the status quo, to prove that listening can be essential to a good discussion, and to challenge politicians and political parties that often cater to talking points and leave so many independent thinkers feeling unheard and unrepresented. Of course, I got some nasty tweets upon my exit from the show as well, but the majority I saw made me feel incredibly grateful.

In several cases, the technology we hold in our hands serves a good purpose. But in a larger sense, as I think about social media in general and my experiences overall, I'm not so sure that the open forum for everyone to consistently and publicly be vocal, opinionated, engaged, commenting, and directing those comments at others via @ signs and direct messages and other tools has been helpful or brought out the best in us, particularly when we encounter things or people we disagree with.

It used to be that television was a one-way street. Television shows came into our living rooms, bringing us comedy or drama or news, and that was that. But now the communication travels both ways. Television shows are broadcast out to viewers, and viewers' reactions can be sent right back to the shows in real time. Whether with phone-in voting shows like *American Idol* or *Dancing with the Stars*, or via Internet communications for *The View* or shows on Fox News or elsewhere, audiences can immediately share their thoughts and feelings about what they see on their screens, including the content, the on-air talent, everything. When you're the voice of dissent or serving up controversial political opinions, positions in which I've been, you get quite a few thoughts and feelings sent your way.

I often get asked if I minded being the minority opinion as the libertarian-conservative on *The View*. I didn't mind it at all. I enjoyed the challenge and the healthy debate it fostered. In fact, I proudly titled my first book *Outnumbered*, which chronicled being a conservative (I would now amend that to libertarian-conservative) in New York City while teaching at a private school, with all of the comedy and discoveries that ensued. I believe in placing yourself in the best arenas you can to change hearts and minds whenever possible, in maximizing your potential to reach people, in engaging with those inclined to disagree with you to break down their potential stereotypes about who you are and what you stand for. If I can get one person every day to say, "Hold on, that actually made sense," or "I didn't realize I'd agree with a 'libertarian-conservative' on some stuff," or "She's not saying what I thought she'd say, that's interesting," then I consider it a good day's work. In fact, when I started at Fox News, I told Roger Ailes, the network's founder and chairman at the time, that I'm not a preach-to-the-choir girl. "I'm going to ruffle feathers on both sides," I said. His response: "That's why I hired you." I appreciated that, and I ruffled plenty.

I have never espoused or modified my views because some TV executive told me it's what they wanted at any given moment. It's just not my style. I also don't jump on board with what everyone around me is doing or thinking. The world doesn't need that. It doesn't need phony talking heads, groupthink, collectivism, or a bunch of ideological robots that don't question the status quo. We need more dissent, more honest debate, more truth tellers, more challenging questions, more diversity of thought and discussion, so that we can all work together to get to the heart of issues and do better for each other.

I used to joke that I popped out of the womb as a Reagan conservative. From an early age, conservative principles resonated with me. But I was challenged quite often by family members who disagreed, by growing up in liberal New York City, by graduating from liberal

Columbia University, and by teaching in a liberal private school. I liked that. Sometimes it made me consider and reconsider my stand on an issue. Other times it made me more passionately defend it. But, most importantly, I learned how to sit in rooms full of people who didn't agree and find our way to respectful debate, a better understanding, and potentially common ground, things you don't learn if you spend your life in an ideological bubble.

I don't often see that desire for respectful debate on social media. I see constant division, name-calling, a my-team-versus-your-team mentality, an inability to reach across the aisle and say, "I see what you're saying, but have you considered this?" We seem to live in a world where we focus on the divisions in the debate rather than the debate itself and take a dig at someone only to rack up retweets. Of course, each of us has times when we passionately stand our ground on an issue, when we fight for what we believe in through robust debates where no one sees eye to eye. But shouldn't we still meet face-to-face when possible, with a willingness to listen, to consider what it's like to walk in someone else's shoes, to understand even if you heartily disagree? Even if we don't meet in person, shouldn't we bring that respect and approach to our online exchanges?

If you've ever participated as a human being in this world in any way, you've likely come to realize that many who disagree with you on policies or perspective share similar concerns, fears, desires, wants, and needs. They want to fix the very same problems you do. They just want to go about it differently. Most people want to do good for themselves and others. I genuinely believe that. I feel that freedom, opportunity, the free market, low taxes, and motivating individuals to fulfill their dreams are a means to generate that goodness. Others may see a different path. But that doesn't necessarily make them bad people. When did we forget that?

#

WHEN I STARTED IN TELEVISION IN 2010, SOCIAL MEDIA WAS happening, of course, but much less so than now. Instagram and Snapchat weren't a thing, Twitter was much newer and significantly tamer. I look at Twitter in 2018 and it barely resembles the Twitter of 2010. It's gotten uglier—a lot uglier. To get retweets, gain followers, and rise to social media fame, people have embraced sound bites, insults, and catchy one-liners that gloss over serious issues. In trying to win some social media popularity contest judged by a bunch of strangers that we have somehow entrusted with making or breaking our egos, smart people who once carefully debated issues now toss out nasty but catchy personal attacks, seemingly without a thought.

The rampant finger-pointing, name-calling, and knee-jerkish culture that thrives in the social media world has made many of us forget what it means to behave like thinking, responsible grown-ups. Sure, this pattern of behavior has built some social media stars, but what has been glorified are the very things people once scolded their children for doing to other kids on the playground. It has often elevated people whose behavior is the very opposite of what we once hoped our kids and grandkids would demonstrate. So many have forgotten our humanity because some guy in Silicon Valley decided to create a medium that rewards thoughtless keyboard courage and empty one-liners. Are we that weak?

Watching our politicians—those elected to represent us—insult each other back and forth on Twitter nauseates me. When it comes to social media, so many political leaders seem to behave like a bunch of tantrum-throwing kindergartners at recess. It's not exclusive to Democrats or Republicans. Politicians of all political persuasions are guilty. The insults they hurl back and forth don't deserve any extra attention, so I won't repeat them here. But as I watch this pattern of behavior unfold time and again, I can't help but wonder:

Aren't those politicians embarrassed to have leaders of other countries witness such childish displays? (I fear I'm insulting children with that comparison.)

How does shouting ugliness back and forth on Twitter make their point?

Which insults help bridge the gap when they have to sit down together in the same room to do the work of writing bills and making laws that the people of this country elected them to do? (Hint: none.)

They *should* be embarrassed, and we should also be embarrassed if we feed their behavior with our "likes" and retweets. A presidential tweet that might seem funny in the moment isn't so funny when the leader of an enemy country reads it and we look like a big joke. It isn't so funny when we have to explain to our kids why the decency we expect from them and each other in our households somehow isn't demanded from some of the most powerful politicians in our country.

It trickles down. There are just too many people, from public to private citizens, who have experienced some form of these attacks online. According to the website DoSomething.org's "Facts About Cyberbullying," which cites eleven different sources, including research studies, data sheets, and articles, "Nearly 43% of kids have been bullied online" and "70% of students report seeing frequent bullying online."

In the article "The Internet Problem We Don't Talk Enough About" by Madeline Buxton, published by Refinery29 on March 15, 2017, the website announced their campaign "Reclaim Your Domain" to "make the internet (and the world outside of it) a safer space for everyone—especially women." The article summarized what they said were "some of the most compelling statistics about online harassment," from sources including the Data & Society Research Institute's report *Online Harassment, Digital Abuse, and Cyberstalking in America* and the Pew Research Center's online harassment study. The findings:

In a survey of over 3,000 Americans, "47% reported experiencing some sort of online harassment or abuse." The figure rose to 65 percent for those eighteen to twenty-nine years old. Sixty-six percent of the harassment was linked to social media sites and apps. Forty-one percent of women ages fifteen to twenty-nine, and 33 percent of men, censor themselves online. And 21 percent have stopped using social media altogether. They also found that "1 in 10 young women has been harassed online by someone threatening to expose nude photos." In summary, Refinery29 observed: "Cyberstalking is turning into an epidemic. An estimated 2.5 million cases occurred over the course of three years."

HATE IS NOTHING NEW. IT'S AS OLD AS THE BIBLE. IN Ecclesiastes—and, of course, put to a great melody in the 1965 Byrds song "Turn! Turn! Turn!"—we are told: "To every thing there is a season . . . A time to love, and a time to hate." Hate and love might even be prehistoric. I saw *Jurassic Park*. Those dinosaurs had their fair share of rage. The reasons, philosophy, and biology of hate would probably fill as many books as the reasons, philosophy, and biology of love. Let's just say that as long as human beings are around, there will be hate and love.

The Internet is not the first medium for sharing biases and agendas for one thing or another, against one thing or another. History is laden with examples of kings, czars, and emperors verbally putting out their messages via town criers and such to the public to rally support for their particular missions. These missives were not so much news as editorials and marketing slogans. There was no pretense that they were objective or inclusive. The language was likely charged and engaging, the energy with which it was announced emotional ("This law will save your life"), encouraging ("Rise up and help us take over those bad guys"), commanding ("New rules, starting today"), or strongly motivating ("If you don't look, act, and pray like everyone else, you might want to get out of town—fast"). Both Eastern and Western cultures used what we now call propaganda to promote their agendas, whether it was to support a new ruler or push through a new law. Not all of the propaganda was propagating something good. There were then, as there are now, people who wanted to send messages of hate, and they did so through whatever medium they could find, to ensure the exclusion, devastation, or eradication of others.

The printing press, invented in the fifteenth century, was probably the biggest boon to hateful, mean messages. Now certain rulers and groups could print up and distribute flyers and pamphlets using the written word, illustrations, graphics, and cartoons to promote their causes. Revolutions, rebellions, imperialism, colonization—all of these things came on the back of the written word. But so did evolutions, reformations, renaissances, and abolitions. As it was then, it is now. There was always love and hate, progress and regression, kindness and cruelty. It's not the machines as much as it is the men who drive the machines. Then as it is now.

Only now we're living in a world where most of us are obsessed with and possessed by the gadgets that connect us. Access to the Internet grows constantly, where the capacity for good is larger and the range of abuses wider. At least it certainly feels that way to me and many others who have been targets of public criticism and nasty comments.

In 2016, Columbia Pictures remade their hit film *Ghostbusters* with an all-female cast that included Kristen Wiig, Melissa McCarthy, Kate McKinnon, and Leslie Jones. According to an account in USA Today College online, "The Whole Leslie Jones Twitter Feud, Explained" by Jamie Altman at Chapman University, "On Monday, July 18th [2016], Jones fell victim to Internet trolls, who posted inflammatory tweets under her account name, made racist comments and sent the actress pornographic images." One of Jones's responses:

"I feel like I'm in a personal hell. I didn't do anything to deserve this. It's just too much. It shouldn't be like this. So hurt right now." Eventually, Jones temporarily took herself off Twitter.

How does a talented and capable actor get pushed off of social media by a bunch of bullies? How does a company like Twitter balance First Amendment rights with a person's right to not be verbally abused? According to USA Today College, "Twitter emphasized that although the social platform is a space for expression and opinions, its rules prohibit targeted abuse." The article included a statement from Twitter that promised, "We are continuing to invest heavily in improving our tools and enforcement systems to better allow us to identify and take faster action on abuse as it's happening and prevent repeat offenders."

Other celebrities came to the defense of Jones, including actor and comedian Amy Schumer. In an August 10, 2016, article in the *Los Angeles Times*, Schumer told writer Glenn Whipp: "Leslie is a friend and we talked about it. It was just so much hate at once. And you feel like it's never-ending and it is the truth. She was not used to it. I've had 10 years of people sending a lot of vitriol and it's been spread out. For her, it hit like a ton of bricks . . . I remember how it feels. It feels like your new reality, like your biggest fear has come true."

On the same note, in a July 12, 2017, CNN online opinion column entitled "Why I'm Breaking Up with Twitter," CNN anchor and former Fox News morning show host Alisyn Camerota addressed Twitter: "Something's happened to you. You're a shadow of your former self, the one I was first attracted to. It's no fun to be with you anymore. You've become mean and verbally abusive. In fact, you gross me out. You're a cesspool of spleen-venting from people who think it's acceptable to insult other people in public and anonymously."

Of course, it's not just actors in films and news anchors on TV who are bullied online. The cruelty can happen to anyone. I found more than a dozen stories of online bullying leading to suicides. In an October 2016 issue of the British paper the *Independent*, the mother of a teenager who killed himself "after years of bullying, much of it exacerbated through internet attacks," wrote an open letter to other parents urging them to guide their children to be responsible and kind. Beyond being saddened that anyone would have to remind parents to teach their children the values of responsibility and kindness, I found it devastating that this child ended his life because, as his mother said, the social media was "cruel and overwhelming."

How painful and sad. All because some nasty cowards hiding behind their computer, phone, or tablet screens decided to gang up on someone in a medium where it's all too easy.

There have always been people who have negative things to say about others. Only now, it seems easier, more public, more visible, and louder. In this era of let's-hear-everything-you-have-to-say-all-the-time media, everyone has a megaphone. And the bad guys really like to use it.

# #

I DISCOVERED THE ANSWER TO MY QUESTION:
How can there be so much hate out there? Or is there?

It was this: *There is and there isn't.* Both hate and love have been around a long time. What we need to do is FEED OUR FRIENDS. If we don't fuel the hate engine, it won't run.

We could all use a lot more KINDNESS online (and in the world in general). Tech devices aren't going anywhere. Neither is social media. We can't control the platform, but we can control ourselves.

When we search the Internet, let's look for the positive things. Let's seek out the positive and uplifting stories. I do that all the time with funny videos or stories of animals being rescued by wonderful people and organizations, and then I retweet those stories to spread those good feelings around.

On a personal level, let's use social media to affirm our connections with, and support, the people we value who are active in our real lives. Let's responsibly remove those from our online presence whom we've already actively removed from our in-person lives. Let's think before we click or type.

We need to act better, too. When we participate on the Internet, let's take a moment to think about how we're saying what we're saying. I'm not saying we should be politically correct, but we should be thoughtful. There's a difference. Political correctness implies sugarcoating, hedging, refusing to speak the truth because someone, somewhere, may get offended. That doesn't interest me. I'm not suggesting we censor uncomfortable facts or stomp on the opinions of insightful commentators and columnists. I'm also not suggesting that we create a culture of overgrown babies who feel triggered by words, are hypersensitive to comedians' jokes, or seek out safe spaces free of colorful humor and challenging ideas. What I *am* suggesting is a thoughtful approach to online mediums—honest, direct, smart, calling things as we see them to get to the core of an issue, but without the personal attacks that can turn anyone into a real-life Mean Girl. We need to think about how we treat people in online conversations and how we'd like them to address us. I'm constantly checking in with myself about this.

Sometimes it's satisfying to find reasons not to like people. Look at all the public anger toward leaders in every arena: government, politics, sports, entertainment, business, media. Too often, we put these folks up on a pedestal only to knock them down. Our technology gives us easy access to do that. Often, when I take a step back, I can find my way to understanding another side of the story. I may not agree, but it humanizes the other viewpoint and hence humanizes my approach. It's so much more rewarding to shake off the tendency to say nasty things. What good does it do us to act so poorly? Even if we think someone else is acting ugly, why should we stoop to their level? A college friend used to often say, "The way you treat others is *your* karma. The way they treat you is *theirs*." I've always believed that the universe rewards good people. It's as good a time as any to remember that.

Let's all use technology to accentuate the positive and try to disempower the negative. Let's feed our friends, not our enemies. Let's foster goodness, and let's behave well ourselves. Don't be the person who must send a midnight message to an ex-boyfriend who you know doesn't want to hear from you. Reconsider before you tweet that mean observation. Check in with yourself when you want to type a few choice words for immediate gratification. Step back a moment before you share or retweet spiteful comments or ideas. You never know what's going on in someone else's life, what struggles they may be facing, and how your words hit them. Don't use your bad day to try to make them have one, too. We can all find a connection, a humanity, with someone who thinks and believes things we don't. Let's use our social media apps to promote the ideas the world needs more of, not less, such as kindness and empathy. I say all this not because I haven't made mistakes with this stuff, but because I have. I made many mistakes on television and social media. There was a time when I'd fall right into the knee-jerk comments and reactions that social media creators wanted me to make—and needed me to make—to empower the medium. Then I woke up and decided to empower myself instead.

What do I do now? I try to ignore or condemn obnoxious tweets from politicians determined to flame-throw to seek attention, rile up a political base, or have a public tantrum they hope will be championed by us. Ignoring tantrum throwers on the playground and on Twitter drives them crazy and is also the best way to make them stop. Attention seekers grow quieter when their bad behavior stops getting attention. I also do my best not to echo that kind of bad social media behavior in my exchanges. Temper tantrums on Twitter aren't a good look for any of us, least of all the people we elect to run our government. If *they* can't do better, *we* sure can.

We have to. Because guess what? It's getting worse. The pattern of behavior that gets rewarded in those mediums has been extended beyond social media and right into—and onto—television news. What was once a platform for respectful debate has too often become a cesspool of bad behavior. Television news has reverted to the Wild West of bad manners and bad characters.

Many television personalities have carried their bad Twitter behavior right on over into bad TV behavior. Some are doing it because they're so used to exhibiting bad behavior online, and spend so much time there, that their on-air behavior thoughtlessly mimics it. Others have seen bad behavior become synonymous with success in others, so they, too, stoop to a level beneath them and act badly. From there, it's a downward spiral. This bad behavior becomes the norm, and so television executives decide that it's what people want, what they crave, what will "make good TV." Now we turn on the television and see commentators and hosts willing—and often eager—to be rude and throw around insulting one-liners and angry sound bites, free of substance and full of the what-can-I-say-to-grab-a-headline? comments that thrive on social media.

Those who try to have a nuanced, real, respectful, honest debate often struggle to figure out how to even exist in the news world now. Real solutions and problem-solving suffer. We all lose out. The way we treat each other and the way younger generations learn from us to treat each other suffer, the dialogue gets degraded to the most infantile exchanges, and many are left feeling agitated and riled up. The worst parts of humanity are being nurtured and rewarded.

It has to stop. But unless we speak up and start changing the behavioral norms in those mediums and beyond, that's the world we're creating, one that elevates words and actions we would've once been embarrassed to condone, let alone support. It's all in our power to change. To change the norms of how we treat each other. To change what standards we demand of the people who hold big microphones on our TVs. To change the channel and let TV executives know that we demand better. Because we deserve better. Television itself deserves better. I like television, I like being on television, I like watching television, I like all of that quality entertainment coming right into my home. I love that my mother gave me this exotic name that she got from a television show. Most of all, I'd like for TV to remain a medium in which I want to engage.

We must act responsibly with the technology we have created so that it gives us what we need *and* honors who we want to be. Thinking before you type, taking a minute to read what you wrote before you angrily hit "send," and saving some of those heartfelt, heated debates for in-person spaces where the dignity of our humanity reigns supreme all serve to make those mediums better spaces for everyone. I've always said about TV that if you're being rude, interrupting, making personal attacks, or shouting your points, you're losing. You're proving that you can't win that debate with a calm, collected, sincere, commonsense approach. Social media and online interactions are no different.

The mediums may have been designed to thrive on shouting from behind anonymous avatars and on rapid-fire, thoughtless exchanges, but it is you and me who decide whether to feed the beast, or to instead cultivate a better representation of humanity.

## CHAPTER 10

LOL (LOSING OUTSIDE LIVES)

What is nature's place on a tech-dominated planet?

WHEN I WAS ABOUT THREE YEARS OLD, MY DAD GOT A phone call. On the landline. In the living room. It was from my mother's sister's husband.

DAD: "Hello?"

UNCLE LARRY: "Tony, it's Larry. Come to Florida."

DAD: "Why?"

LARRY: "It's warm. And there's a job here for you."

DAD: "I do like warm. But I have a job."

LARRY: "The coins and stamps will wait."

My dad is a philatelist and numismatist. If you know what those are, send me a note so I can send you back a gold star. I have yet to meet anyone who has any idea what I'm talking about until I explain that my father works in a stamp and coin business and used to collect those things.

DAD: "What's the job?"
LARRY: "That hotel I built. You know, the one on the beach in Fort Myers? It needs a manager. Joan and I are going to live there and Linda and you and the kid [that was me] can, too."

My dad turned to my mom. "Larry wants us to move to Florida so I can manage a hotel."

MOM: "Is he nuts? Tell him there's sun in Florida, and sand, and humidity." [Mom wasn't a beach girl by any stretch of the imagination.]
DAD: "I think he knows all that. That's the point."
MOM: "We can't leave the house. And my parents and our friends are here."

We could all hear Uncle Larry through the receiver: "Tell Linda they can all come down whenever they want. I can block out several rooms for them."

My dad looked at my mom and shrugged. He hung up the phone, and he and Mom talked and talked, and then talked some more. Dad always loved Florida. Unlike Mom, who gravitated toward New England getaways and Manhattan's Museum Mile, Dad was a year-round-tan, beach-town kind of guy. Mom was hesitant to leave the quaint condo she loved that was close to Brooklyn and Manhattan, where family and friends resided. But an adventure on the Gulf of Mexico was an opportunity she couldn't pass up. They said yes and started making arrangements for our move.

The hotel was right on the beach in Fort Myers, Florida, a (then and now) small town on the Gulf Coast—the calmer, quieter, less touristy west side of Florida. In the 1980s, this was simple living at its best. In our immediate neighborhood there were the hotel, some motels and beach condos, plenty of white, fluffy sand, and a long pier from which fishermen cast their lines in the morning and couples strolled hand in hand at night. There were a few hamburger shacks, ice cream shops, and some stores that sold colorful towels, wide-brim hats, plastic alligators, and all things beachy. There was also a 7-Eleven and a Dairy Queen I would come to know all too well.

As promised, we all had rooms in the hotel. Mom, Dad, and I lived on the top floor in a two-bedroom suite. Before you start thinking all kinds of fancy things that go along with the word "suite," the hotel was a small, five-story beach hotel. It was cozy, adorable, and well kept, but it wasn't fancy. My cousin Joey, who was twenty-six at the time, lived on the first floor. Aunt Joan and Uncle Larry were on the second. My mom's good friend, Rose, who was like an aunt to me, came down soon after we arrived to manage the hotel cleaning crew and live on the third floor. My mom's parents, Nanny and Poppy, who you've gotten to know a bit by now from my Brooklyn stories, came down for the winter months and other special occasions, and would stay in different rooms each time.

We would all often have our breakfast on the small hotel balconies that accompanied each room. Mom and Aunt Joan would shout out to each other, and to the others, the schedule for the day. It was old-school, multigenerational living at its best. Our living room was always filled with parents, grandparents, aunts—the whole crew. I can remember huddling together in our suite during hurricane season. One time, the first few floors of the hotel flooded and the whole family moved up to the fifth floor for a week. We spent a lot of time together then, in our suite. We ate lots of homemade Italian food, watched vigorous winds challenge the sturdiness of palm trees, and lost power completely for several days. We played board games and talked for hours by candlelight at night.

It's funny, because my friends often say that they feel transitory and unsettled when they stay in hotels. I get it. I mean, you're usually living out of a suitcase in a space that's not your own, sleeping in a bed that a total stranger slept in a short time before you got there, the same bed another total stranger will sleep in soon after you leave. It's an odd concept if you think about it. But for me, because I lived in a hotel for a couple of years as a tiny tot, hotels always feel kind of homey.

Dad was a no-nonsense, the-training-wheels-are-off kind of parent. He wanted me to be brave right from the start. So, at three, my water-wing floaties came off, and I was swimming the full length of the hotel pool and back, unassisted and underwater. Mom would make sure I had all my sunscreen on, and I'd often hear her shouting from the balcony, "Tony! Don't let her swim underwater too long!" During the day, when I wasn't in school, we were on the beach. We collected seashells, played shuffleboard, made sandcastles (and snowmen out of sand in the winter), flew kites, and grabbed snacks at the snack shack nearby. For my fourth birthday, Mom had actors dressed as Snow White and the Seven Dwarfs appear at my beach party. Snow White was my absolute favorite. At night, we played card games, talked, cooked, ate, and ate some more. When Poppy came to visit, he'd walk with me every night to the 7-Eleven up the street, and I'd pick out my favorite snack. A few times a week, Dad would take me to the local Dairy Queen for a soft-serve vanilla ice cream cone with rainbow sprinkles. Mom and Aunt Joan took me for lunch every day after school at what we called the "hangout mall." Mom says I'd spill my soda almost every day by accident somehow. I spilled it so often that the café owner would get the mop ready when we walked in.

Eventually, Mom missed home. She missed the changing of the seasons, our condo, and our family and friends. Hotel ownership was shifting. Two years in Florida had been an amazing adventure, but it was time to head back to Staten Island. We had rented out our Staten Island condo for the two years we were away, so when we returned, I was able to head back to my old room. Dad resumed his work with coins and stamps, Mom got a job managing a clothing store, and I found my way, at the ripe old age of five, back into the real world of first grade.

#

THE MEMORIES OF THOSE DAYS IN FLORIDA ARE PARTICU-larly vivid for me. Everyone was very much there, engaged, doing things together. We were outside in the elements, enjoying the fresh air, the cool sea breeze, the Florida flora and fauna. Birds swooped, fish swam, salamanders slithered. We sat in beach chairs, bare feet in the sand, talking, or not talking, present with and aware of each other. We flew kites in the wind, grilled delicious food outside on the beach, swam in the ocean. As with the virtual reality rides at Universal Studios, all of my senses were activated. I could feel the sun on my skin; see, smell, and taste the fresh, salty air; hear the laughter; absorb the joy of being with those most familiar, *la famiglia*.

Nowadays I walk on the beach and what do I see? Almost everyone on their phones or their tablets. Including me. Even I've had days at the beach taking, taking, taking pictures and then posting, posting, posting them to Instagram.

We've been hijacked, pulled from the elements of earth, wind, fire, and water, and dropped into a box of four walls and a roof filled with tech boxes. Even when outside, we're consumed with one tech box or another. It's like we've been programmed to be obsessively plugged into some mechanical device at the expense of experiencing the sand, the ocean, the simplicity of flying a kite without rushing to get the perfect, filtered, photoshopped picture of us flying it. I've talked about this with my friends, relatives, and colleagues, checking in to see if any of this bothers them, too.

The responses go something like this:

"Outside? My kids don't go outside."

"Too much tech? You're telling me."

"Oh, I know, isn't it awful?"

"I hate it."

"It's bad, AJ. I see it with my sister. She's on that thing all the time—hardly hears a word I say. We visited our friends in the country at this beautiful ranch, and she was head down, on her phone the whole weekend."

"He's constantly texting. Every time I see him, he's swiping at his phone."

"Have you been to a park lately? No one says hi anymore. While the kids play, all the parents just stare down at their phones, even if it's a beautiful day outside."

"Forget vacations. It's impossible to get my kids to leave the phones and iPads in the hotel room and just enjoy the pool or play some beach volleyball."

Okay, so we realize what's happening. We get it. But why do we accept it as some unavoidable next stage of humanity? Why don't we all make a change and hang out with the people we enjoy on the beach, at a park, in the backyard, or on a city stoop while actually making eye contact with them, free from the distraction of this phone buzzing or that person Snapchatting the whole thing or someone else trying to capture the right moment for Instagram? The outside world, with its magical vistas, gorgeous sunrises, forests, rivers, and mountains, is there for us to dive into and enjoy. Even if you live in a city, there are parks, promenades, and boardwalks to explore.

Yet, now the outside world is being infiltrated by tech. As was the case in 2016, when Pokémon took over the planet. As you may recall, Pokémon was the popular video and card game franchise from the 1990s in which players, called Trainers, would collect various types/species of Pokémon characters to compete against each other. Okay, I never quite understood it, either, but there it is. In a July 16, 2016, article in the *Economist* called "I Mug You, Pikachu!," the subtitle observed, "A Hit Video Game Shows How the Real and Virtual Worlds Are Merging." This was referencing the phenomenon that was the July 6, 2016, release of Pokémon GO, a Nintendo app for smartphones. This app put the Pokémon game right into the real world, one of the first video games to require players to go outside of their houses to play. In a July 13, 2016, article on the website Bustle titled "How Does 'Pokemon Go' Work? Here's Everything We Know About the Tech Behind the Augmented Reality Fad," Claire Warner explained: "The game uses your phone's GPS, camera, and clock to generate a brightly-colored, fantastical version of the real world on your phone. Certain Pokemon tend to be found in their 'natural' habitats; if you're walking by a lake, for example, you're more likely to find a Squirtle or Poliwag. Players have found Pokemon pretty much anywhere—the street, local landmarks, and even in the middle of gigantic fountains."

Participants would look into their phone's camera feature, see the real world, and find animated Pokémon characters here and there in that real world—on a biking trail they were riding, across a crowded city street—mixing reality and animation right there in the palm of their hands. Soon enough, participants were holding up their phones, spotting Pokémon characters, and chasing those character images all over the place . . . in parks, museums, even cemeteries. The game was played everywhere.

That wasn't always a good thing.

A 2016 article on the JAMA (*Journal of the American Medical Association*) Network called "Pokémon GO—A New Distraction for Drivers and Pedestrians" noted the dangers of pulling game characters off the page and putting them onto life's stage. "Pokémon GO, an augmented reality game, has swept the nation," they wrote. "As players move, their avatar moves within the game, and players are then rewarded for collecting Pokémon placed in real-world locations. By rewarding movement, the game incentivizes physical activity. However, if players use their cars to search for Pokémon they negate any health benefit and incur serious risk."

The article added, "Motor vehicle crashes are the leading cause of death among 16- to 24-year-olds, whom the game targets. Moreover, according to the American Automobile Association, 59% of all crashes among young drivers involve distractions within 6 seconds of the accident. We report on an assessment of drivers and pedestrians distracted by Pokémon GO and crashes potentially caused by Pokémon GO by mining social and news media reports."

The virtual invades life. Life becomes about the virtual.

My nostalgia for kickball and freeze tag was getting richer by the day.

What about going to museums or art galleries? Are those excursions untouched by tech? Nope. In July of 2017, a video went viral when a woman stumbled into a piece of art at a place called the 14th Factory in the Lincoln Heights area of Los Angeles. The pop-up art exhibit had a room with dozens of sculptural crowns on a series of white columns, where visitors were encouraged to take pictures of themselves—selfies—with the art. The patron stepped a bit too close to one of the columns and fell back into it. According to a July 14, 2017, report on Fortune.com, "That stumble and the domino-like effect it started permanently damaged three sculptures and caused varying degrees to several others . . . The approximate total cost of those damages came in at $200,000."

Then there's the all-American game of baseball. Nothing harkens back to the days of quiet, simple, slow-paced life quite like baseball. Snack and refreshment in hand, you sit outside, at least at most Major League Baseball parks, and enjoy watching athletes engage in a nice, measured game of hits and outs, one of the few that doesn't even bother, for the most part, with time, or even a clock. The crowd enjoys the stillness between moments, the quiet anticipation of what's to come. Although the professional game has been tarnished by scandals for decades, many of us still hold out hope that the nation's favorite pastime could be the last vestige of old-fashioned goodness, of being present and in the moment, immune to the character-shaking effects of high-tech . . .

Not so fast. On September 5, 2017, the *New York Times* reported that the Boston Red Sox "used an Apple Watch to gain an advantage against the Yankees and other teams . . . [and that they] executed a scheme to illicitly steal hand signals from opponents' catchers . . ."

Honestly, baseball? What's next, apple pie? Instead of savoring the crust and fruit, cinnamon and sugar, will someone invent an app that sends a signal to your brain to make you think you already ate it? Don't put it past them. Seriously, is nothing sacred anymore, untouched by the hand of technology? Untouched, not because no one thought of a trendy new invention, but because we, the consumers, finally said, "I don't want my baseball messed with!"

I was desperate to find an activity, a space, a gathering of people that hadn't been invaded by technology. So I hopped in my figurative DeLorean . . .

# #

CYNDI LAUPER IS RIGHT—SOMETIMES GIRLS JUST WANT TO have fun. One Saturday I decided I wanted to unplug completely, get out of the apartment, and dance the night away. What better way to do that than to head over to a big warehouse space in Brooklyn and step into the 1990s. A retro party. Absolutely perfect.

The huge space was filled with mostly twenty-, thirty-, and forty-somethings, a majority of whom were dressed simply in jeans and flat, easy-to-dance-in shoes, bouncing about, singing to Nirvana, Madonna, the Spice Girls, Oasis, Britney Spears, and other oldies but goodies. I scanned the crowd. It was a bunch of people having a good time. Everyone was engaged with the people around them in a really fun way. There's something about singing the lyrics to great songs collectively, when you all know the words, that makes things infinitely more fun.

In the three hours I was there, I think I saw two phones out, both to shoot videos. But the bright lights of the snooping, conspicuous cameras took those around them out of the moment, and after several annoyed reactions, as well as the quick realization of the phones' owners that they were messing with the energy of the moment, both of those phones were put away pretty quickly. I didn't see any selfies, no primping or posing or staging, no trying to capture some moment to alert the world to an outfit or hairstyle or presence. There was a bizarre attention to the people, music, and energy right there in the room, an inattention to whatever was going on outside that space.

I use the word "bizarre" because it was such an unbelievable contrast to the nights out that many experience on the regular now, where the bar or lounge is littered with technology—everyone constantly on their phones, connected to an app or inside their social media, swiping left and right, eager to participate in the hookup culture, the room filled with men and women constantly adjusting their hair or wardrobe for the perfect selfie, a bunch of people talking to the people inside their devices rather than the ones right in front of them, a sea of disconnected faces. Although I love to dance, I'm not into the New York City lounge scene these days, because the mood is often aloof, self-absorbed, and device-centric.

In contrast, the people who had chosen to attend the 1990s party I went to weren't in that head at all. I wondered if they had all felt the need for an unplugged night like I had and gravitated to a retro party in search of something they remembered and longed to regain.

Then it hit me.

If enough of us felt this way, were self-aware, and acted upon it, the spaces would change, the mood would change, people's connections to real-life moments and the people in front of them in those spaces would grow. The norm would change. The Manhattan lounge scene—and so many other spaces in so many other places—would transform.

We don't need to hop into a DeLorean or rely on a retro escape party. We can create that old-time feeling ourselves, anywhere and anytime we want, by simply being plugged into real life and un-plugged from tech distractions. If we adjust our behavior, one person at a time, the whole room can shift, the entire vibe can change, life can change.

At the '90s party, there was an old-fashioned photo booth, one of those tiny ones where you step in, sit down, pull the curtain closed, and the camera snaps strips of four square photos. It was fascinating to me that in a room full of people in 2017, many of whom likely had smartphones, I didn't see one selfie being taken, yet there was a perpetual line for that old-school photo booth. I wondered if, in the midst of that crowd and that music and that mood, they had forgotten about their phones altogether. Or maybe they wanted the experience of squishing inside the booth with friends or a significant other, and the hilarity of it all. Or maybe it was something about those tiny black-and-white photos that was reminiscent of something that made them feel a certain way, a way they liked. Regardless, it seemed as though once everyone was feeling the 1990s groove, smartphones had no place there.

This whole thing drives me crazy, though. Because all it takes is our decision, one at a time and then collectively, to take back what's been lost. It just takes a few minutes to think about it all, to recall what it was like to engage as human beings and to create the moments to do so.

These can be the choices we make. Yet, we don't. Why not? Nowadays, too many of us don't know what to do without a cell phone in hand, or at least nearby. Including me.

One day, I decided to go for a walk without my cell phone, to observe the outside world. I noticed a kid on a bike learning to balance without training wheels. I remembered my childhood doing the same. I thought of my dad, who put my training wheels on and soon took them off when he knew I was ready. Filled with that memory, and others it triggered, I walked with a lightness in my step and looked up. I noticed that the sky wasn't just blue, but various shades of light blue, dark blue, purple, indigo, pink, peach, orange, and white. I saw a guy painting his brownstone door red. I thought about how much I loved red doors. And picket fences and shutters. I fell into a stream of consciousness that led to me imagining what it would be like to have my own house, with a garage and everything, and from there my mind was off and running, building the best life I could for myself.

Maybe if I took more time to look at those squirrels that just sprinted past, I'd be amazed at how fast they can climb a tree and jump through the air from branch to branch. I'd think about their claws, how their feet rotate, and wonder at the amazing technology that nature had created all on its own. From there I might come up with the next best idea.

Spending time in nature, outside in the world, is good for us. Spending too much time inside, closed up, under artificial lights, glued to devices, is not.

Historically, when we were farming and cultivating the land and laying bricks and all that, we spent a lot of time outside. Then, as countries entered the industrial age, more work moved indoors. Now, in the computer age, many of us spend a large amount of time indoors. Various studies claim that the time we spend outside ranges from 5 to 7 percent a day, and some of that is just getting here and there, not sitting outside, enjoying the sky, the trees, the ocean, or the stars.

According to Dr. Stephen Ilardi, in an October 26, 2017, *Wall Street Journal* article titled "Why Personal Tech Is Depressing," we not only need to spend more time outside and off of our computers, but we need physical activity and action. If we don't, we tend to have a greater chance of being in a dark funk: "Our indoor, sedentary and socially isolated lives leave us vulnerable to depression." He further explained the importance of physical activity: "Labor-saving inventions, from the Roomba to Netflix, spare us the arduous tasks of our grandparents' generation. But small actions like vacuuming and returning videotapes can have a positive impact on our well-being. Even modest physical activity can mitigate stress and stimulate the brain's release of dopamine and serotonin—powerful neurotransmitters that help spark motivation and regulate emotions. Remove physical exertion, and our brain's pleasure centers can go dormant . . ."

Ilardi added, "In theory, labor-saving apps and automation create free time that we could use to hit the beach or join a kickball league. But that isn't what tends to happen. We're wired, like our ancestors, to conserve energy whenever possible—to be lazy when no exertion is required—an evolutionary explanation for your tendency to sit around after work. Excessive screen time lulls us ever deeper into habitual inactivity, overstimulates the nervous system and increases production of the stress hormone cortisol. In the short term, cortisol helps us react to high-pressure situations, but when chronically activated, it triggers the brain's toxic runaway stress response, which researchers have identified as an ultimate driver of depressive illness."

Here's the key part: "A few generations ago, people spent most of their waking hours outdoors. Direct sunlight boosts the brain's serotonin circuitry, protects against seasonal affective disorder and triggers the eyes' light receptors, which regulate the body's internal clock and sleep patterns—yet we spend 93% of our time inside. Our mood suffers, and our body loses the ability to find restorative sleep. And bathing our eyes in artificial lighting—especially the blueshifted hues of flat screens—stalls the body's nightly release of melatonin, the drowsiness-inducing hormone, until 45 minutes after we power down. The resulting sleep deprivation can both trigger and compound depression."

It's just messy all around. We all need to get outside. Whenever I do, my body and mind thank me.

#

A MARCH 25, 2016, HEADLINE ON TIME.COM REVEALED, "U.K. Kids Spend Less Time Outside Than Prison Inmates, Study Says." The article said that "a survey of 2,000 parents in [the] United Kingdom finds that nearly three quarters of children are spending less than one hour outside every day . . . [and about] 20% are not going outside at all on a regular basis." Meanwhile, according to the article, the United Nations mandates that prisoners must spend at least an hour every day exercising outside.

Who could've imagined that there would ever be a day when prisoners, people in jail, would spend more time outside than children, who are free? Why would kids stay inside when they could be outside playing while breathing in fresh air? We all know the answer. It's because these days they would rather be digging into their devices than digging into the dirt. Writer Richard Louv coined the phrase "nature deficit disorder" in his 2005 book, *Last Child in the Woods*, arguing that there is a cost to raising kids who have no relationship with the natural world.

This has me asking a lot of questions. What does it do to a child—to a person—to not understand what it is to play outside? How can we build character if we don't skin our knees sliding into third base, scratch our elbows falling off our bikes, bump our heads on the next branch of that tree? How can we appreciate the earth, the very idea of being a natural, living, breathing organism, if we don't have a relationship with nature, with the very life-giving systems that exist outside our windows?

It's not just kids. It's all of us. Many of us spend most of our time inside, on our devices, our brains focused in on the keyboard, the screen, even the VR glasses. What does that do to our character, values, and priorities? To our communication skills and interpersonal relationships? To our stress levels? To our conscience and capacity to act like, to be, a human being? Why have we allowed so much of our lives, our minds, to be guided by the fake, inside, virtual world?

Do we know we have allowed it? Or are we at best blindly tapping our devices, accepting the new normal without pause or question, at worst hypnotized by the latest and greatest bright shiny things pendulating before our eyes?

But wait, there's more. Apparently, not-so-good changes in us are happening whether we are *actually using* our devices or not.

According to the April 2017 *Journal of the Association for Consumer Research*, published by the University of Chicago, there was a study that found—are you ready for this?—that just having our devices around, in our hands, in our pockets, sitting on the table, anywhere nearby, can limit our ability to think clearly. In the article "Brain Drain: The Mere Presence of One's Own Smartphone Reduces Available Cognitive Capacity," the authors say that even people who are focused on what they are doing, who are *not* checking their phones, are being affected by the fact that their phones are even near them. The mere presence of phones nearby, in our sight, is distracting. Talk about secondhand smoke—our gadgets are invasive and disruptive when we use them *and* when we don't.

# #

I THOUGHT ABOUT MY QUESTION:
What is nature's place on a tech-dominated planet?

My answer was: *It's the source of everything we are, and we need to get back to it.* We need to step outside.

It's time to NURTURE OURSELVES WITH SOME NATURE. Perhaps we've always, as Henry David Thoreau put it in his book *Walden*, led "lives of quiet desperation," but it seems now, in these noisy times, that we are, not so quietly, more desperate than ever. We need to get to Thoreau's pond, sit on its banks, pull at the grass, and watch the breeze dance through the tree branches.

Tech immersion is a poor replacement for a great conversation on a backyard deck, a solo walk absorbing the sounds of the neighbor-hood, catching a sunset, cooking a delicious meal over a grill in the backyard, taking the family for a swim at the lake.

These days, I get outside a lot more. Bike rides, promenade walks, enjoying outdoor cafés with the people I love most—you name it. Even while writing this book, I took a lot of breaks to breathe the outside air and let it remind me what I was fighting for.

For me, real life will always come first. That's why I feel so lucky to have found someone else who feels the same way. Remember when I told you about my parents' and grandparents' romances and how I started to doubt the possibility of old-school romance in this tech-dominated world? How I said that if I wanted old-fashioned romance, I'd have to go old?

Well, let's just say it didn't quite go that way.

## CHAPTER 11

SUNSETS ON THE ROOF

Who has time for love?

JEREMY AND I MET AT A GYM. WE WERE BOTH IN A PERIOD of fitness exploration. Working out has always been a source of stress relief for me, and I thought I might get into new areas of training to keep it fun. He was in a similar head. He had even taken it a step further and joined a fitness competition as an experiment to test his discipline and willpower. We had the same trainer, who thought we should get to know each other, and so he introduced us. Not by phone, not by text, not by email, but right there in person, surrounded by the sounds of weights clanking and cycles spinning.

We talked a little bit that day about fitness, not much more. But each time we saw each other in the gym after that, we'd talk about other stuff, too. Vacations, work, that guy a few blocks away who'd dress up on any given day like it was Halloween and walk the streets with his hairless cat. Like I said earlier, Manhattan never disappoints.

At first, we didn't have each other's phone number. Secretly, I think we each purposely avoided the classic exchange of digits. There was something special about keeping those first contacts in person. I'd leave the gym with "Hey, I'll be here Wednesday at five. See you soon." Wednesday at 5:00 p.m. would roll around, and he'd show up with a smile, hit the weights, and hang out afterward to talk about whatever we felt like sharing. I remember that he took a trip to Italy in those first few weeks and brought me back a bag of pasta with little Italian flags on the noodles. I thought that was pretty cute. And delicious.

Jeremy was an interesting guy. He was in his twenties but wasn't into the typical Manhattan twenty-something scene by the time I met him. He had dabbled in the club scene right out of college but decided it wasn't for him. He was the kind of guy who liked get-togethers at friends' homes or playing sports, preferably outdoors. He also really liked a good conversation.

In one of our gym talks about social media, I learned that he wasn't into the whole thing. He had started using it in high school and college, but found almost all of it to be a time sink that would eat up minutes, hours, unnecessarily. Before we met, he had disabled his Facebook page because he hated the constant beckoning for distraction. (He later deleted it altogether.) He had started a Twitter account a few years back to try to get concert tickets by promoting festivals, but wound up posting for only one or two events and never wanted to use it to interact with anyone, so he ditched it. He did have an Instagram account when we met (since deleted) but was struggling with what to do about it from one of our first conversations. He had used it here and there to post some photos, but didn't like the idea it promoted of adding more and more "followers" to form superficial relationships at best and engage total strangers at worst. Instead, he wanted to focus on quality time in real life with the people he cared about.

"Why am I looking at some guy I barely know's pool party pic, followed by a selfie of a girl I went to school with years ago and don't even like?"

Our trainer suggested he keep his Instagram account to post

some fitness-related photos through the competition, so he did. He even tried to be open-minded about the app and posted a few pictures of us when we started dating. But soon after, he decided it was time to hit "delete." He had tried Snapchat for a little bit in college, but liked that one the least, so he pretty quickly deleted the app, forgot about it, and later deleted the account altogether when he was irritated to find that deleting the app left the account inactive but still intact. He had also tried a dating app a few times post-college but quickly decided it wasn't for him. If he was going to meet someone, he was going to do it the old-fashioned way.

He was a very interesting twenty-something for Manhattan's twenty-first-century scene.

About two months went by before we exchanged digits and talked outside the gym. We met up for burgers. We sat at a small table and talked for a few hours. The conversation was engaging, with both of us present, no phones in sight. We laughed a lot. That was June. In July we decided to take a walk in Central Park. We stopped by a smoothie bar, then began what would become a four-hour stroll. We talked and talked and talked some more. In August we had our first real date. He picked me up at my apartment, which happened to be just a block or so away from his. We walked over to the High Line, an elevated trail that runs almost one and a half miles along an old railroad track on the west side of Manhattan, what they now call an "infrastructure reuse project."

I'm a fan of the High Line when it's not too crowded. The trail is a long, narrow, curving slice of nature, built around remnants of old railroad ties and tracks, with margins of lush vegetation and rows of bright flowers framed in metal and wood. Walkers travel on pebbled cement and gravel surfaces around and through buildings and enjoy relaxing, inviting sundecks and unique views of skyscrapers and the Hudson River. Plus, it's a top spot to people-watch.

We climbed up the stairs at its northern end on Thirty-Fourth Street and made our way south, high above Chelsea and the

Meatpacking District to where it ends in Greenwich Village. We walked, we talked, we sat on a bench, took in the views of the city, the river, people in their office and hotel windows, strolled through an outdoor sculpture exhibit, talked some more, got some snacks at a brightly colored food stand, and continued walking and talking until we reached West Twelfth Street, where we climbed back down the stairs. Then we walked north again through the city, stopped to pick up some food, and headed back to my building's rooftop to talk some more. That's when I realized that we had been walking and talking for about seven hours. I remember thinking how refreshing it was that we hadn't done anything else except talk.

I oddly felt like I'd known him forever.

At that time, I was only one year removed from my awful experience with Kyle, the Cell Phone Con Man, and hadn't dated at all since. I wasn't even sure what the interaction with Jeremy was at first. He was younger than me, which was something very different for my life, for sure. He had been flirting in the most adorable of ways for some time now, and I was getting to like him more and more, but I was also taking it slow, as I frustratingly tend to do.

"I'm such a magnet for bad guys," I said as we landed on my building's rooftop.

"Maybe you don't pay attention to the good guy right next to you who wants to date you," he countered.

I definitely heard the comment, definitely smiled, and definitely changed the subject to something utterly ridiculous about almond butter or city architecture, because I had absolutely no idea what to say.

But I'm pretty sure he saw me smile.

We sat on two lounge chairs that faced each other, looked at the stars above, the city lights below, and talked for another few hours until around 1:00 a.m. It seemed like neither of us wanted that night to end.

I talked about small towns and how I had always felt out of place in the big city. He talked about Houston, where he had gone to college, and about some of the surrounding towns he thought I'd love and he'd like to show me.

At one point there was a pause, and we just looked at each other for what seemed like forever but was likely about ten seconds. He sighed and said, "I want to kiss you, but you moved your chair and there's all this distance between us!"

I laughed. Jeremy was that kind of guy. Honest, open, silly, the kind of guy who would tell you what he was thinking no matter how it came out.

I loved it.

"Not yet," I said. "I'm still getting to know you."

He smiled and nodded in an intrigued kind of way.

He walked me back to my apartment, said good night, and went home. We both went to sleep thinking about the great day we had had together. It was just right.

To my surprise and delight, all of the dates were like that. The next time we met up, we took a cab over to Brooklyn, then walked back to Manhattan over the Brooklyn Bridge. The first steel-wire suspension bridge ever constructed, the Brooklyn Bridge is one of the most iconic landmarks in the city. Locals and tourists alike are usually quite taken with its limestone and granite façades, soaring towers, Gothic arches, and stunning views of all five boroughs up and down the East River. Jeremy loved it as much as I did. That day was another one full of conversation and connecting, walking and talking and taking it all in, side by side, for hours.

The next date was the East River Greenway, a Manhattan walkway up and down the East River, and . . .

You get the gist.

Over the next several months, we hit all of the walkable landmarks, the unknown nuggets, the undiscovered nooks and crannies that make New York City an explorer's paradise.

No phones, no texting, no nothing, just us.

We were in the world, we were in a conversation, and we were falling fast in love.

In the real world.

I had somehow escaped the tech takeover and run into romance head-on, just like my parents and grandparents did. And with a low-tech millennial, to boot.

Jeremy was the anti-Kyle.

I KNEW JEREMY WAS THE MAN FOR ME WHEN WE TOOK OUR first trip together a few months into dating, and he put his phone on airplane mode and stuck it in a drawer for three days. Still harboring anxiety from my horrifying experience with Kyle, I assumed he did that to hide some life-altering bad habit or behavior. I soon realized that, no, he just didn't want the interruption. I knew that because when he took the phone back off airplane mode, he was leaning on my shoulder, scrolling through the few emails or messages that came in delayed, with me right there, head against his. Not to prove a point, just because that's how he lived. He wasn't trying to hide anything.

Time passed and we got to know each other better and better. One weekend in the middle of June in 2017, Jeremy suggested we escape from work, from the city, from the social scene, from technology. "Let's go to the country. We'll go biking, spend time in nature."

Perfect.

One of the reasons he and I click so well is that we both seem to have been born in the wrong generation. We like the simple things. Good food, good music, long walks. We enjoy others' company without a demanding device present. We talk about anything and everything.

That Friday afternoon, we hopped in a rental car and made our way out of the city. The last of the steel structures receded in our rearview mirror, and we continued south down the tree-lined highways. We talked for a bit, debriefed our days. Then we looked for some music to accompany the scenery and our transition into calm and repose. It wasn't easy. We pushed the arrows up and down the frequencies and couldn't find a good radio station to listen to. One song after another was a singer punching out words about money and fame and women doing this and that for him, and this and that to him, so we turned it off. We knew this was par for the music course these days, and that these ideas had become so normalized that it wasn't uncommon to hear ten-year-olds singing along with the radio about their love for big bottoms, booty, and the shape of your body, but come on. It was so ridiculous that it almost made you laugh. Only, when you thought about it, it wasn't so funny.

It's so interesting how you grow up singing song lyrics without even thinking about what you're saying, or how what you're saying impacts how you view the world, for better or worse. Your perception of relationships, self-esteem, gender roles, family, career, what defines happiness—so much of it is shaped by the music you grow up with. You parrot the lyrics without thinking. I know I did. Until one day something happens in your life and you start to look at the world and people and patterns of behavior with concern, wondering how it all got that way. Those lyrics don't seem so harmless anymore.

Back to our car ride.

That particular day, Jeremy and I opted to shut off the radio, enjoy nature, and share some old stories. The drive took about ninety minutes down to our destination in New Hope, Pennsylvania. Jeremy had found a simple and beautiful bed-and-breakfast. It was small, not too many guests at one time, chickens on the property, fresh eggs every morning. It was my farm fantasy. If I had fifty million dollars, this would still be a place I'd want to vacation, surrounded by endless trees, wildflowers, simplicity, and serenity.

We had a great afternoon of he-and-me time. The next day, too.

Biking, exploring the town, swimming. That Saturday night, we had a dinner reservation at 7:30. Jeremy was, surprisingly, dressed early, which never happened, and was sitting in a chair, watching me finish getting ready.

"Why are you ready so early?" I asked. Typically he'd drive me a little nuts by waiting until I was fully dressed and halfway out the door, only then to start putting on his belt, shoes, et cetera.

But this night was different.

"I just thought I'd be ready to go for a change."

We headed downstairs and he led me a different way, into a small, enclosed garden in the back, where there was a bottle of wine on a table near one of those little fountains that make soothing water sounds. The garden was lit with tiny white Christmas-style lights. He talked a little about me, about us, and then there he was on his knee.

I said yes.

The simplicity of the proposal, amid nature and quiet, told me he knew me better than I had realized. He later told me that he had contemplated doing it on television, but after much back-and-forth with producers, he decided that the whole thing wasn't me. He was right. I was much more of a country garden kind of girl.

We went to an adorable little Italian restaurant. The whole place sat no more than twenty-five people. In one room there was a woman with her son and his girlfriend. Behind them, a table with two couples. In another room, one big family. The whole place had the cozy feeling of candlelight, conversation, and intimacy. I looked around once more and realized—there wasn't a cell phone in sight. Everyone in the restaurant was engaged with those who were there with them. There was none of that plugged-in feeling I get in the city. It was all very present, and real, and delightful. The food was homemade. The owner came out and talked to us—to everyone. He reminded me of my grandpa. His marinara sauce was almost as good as Poppy's. I said *almost*.

The night was beautiful.

The next day, on our way out of town, Jeremy surprised me again. "I want to take you to see a musical."

Whoa. Glitch. Foot on the brake. He put a ring on it and now—this? I thought he knew me. My mom directs musicals for children. I've seen many of her wonderful shows (and others) and love watching her work come alive on stage, but Jeremy knows I'm more of a regular play kind of gal. Preferably a small, original comedy or drama, off-Broadway. Call me crazy, but I always find that the singing and dancing somehow get in the way of the words.

"A musical?" I said, confused.

He grinned. "You'll like this one, I promise."

We packed up our bags, left the inn, and went to the Bucks County Playhouse to see a musical called *Buddy: The Buddy Holly Story.* The theater was a big old beautiful barn on the site of a former mill on the bank of the Delaware River. I loved it immediately.

Whoa. Put that foot back on the gas pedal. He knew me, he knows me, this is my guy.

We walked in and the first thing I saw was a banner that read: "School Dance, 1959." Jeremy had done it. He'd merged my '50s and my farm fantasies into one weekend.

I couldn't believe the theater and the show. I was in my De-Lorean daydream. The characters were singing wholesome, 1950s good-times music. The story was about a young musician, the real-life rock star Buddy Holly, from Lubbock, Texas, who dreams of opening for Elvis and does so less than a year after graduating high school. Soon he and his band are a hit in their own right. He ends up eloping with a girl he asked to marry him on their first date. The romance and the dream are both short-lived, though, as Buddy and two other famous singers, Ritchie Valens and J. P. "The Big Bopper" Richardson, end up dying tragically in a plane crash. It's a real story, a real romance, a real tragedy. But it's still a beautiful tale about falling in love with music, with a dream, with a girl. The settings were a dance hall, an ice cream parlor, and a high school radio station. The characters were dancing and singing and talking directly to each other, no phones or tablets in sight. No hypersexualized music degrading women. No songs about bumping and grinding, loveless sex, or glorification of cheating and drugs. These songs had words that shared a simplicity with the setting that surrounded them, songs about love and innocence and romance. The audience was full of mostly older couples who were singing and dancing along, with a sprinkling of younger people here and there.

During the curtain call, I leaned over and whispered to Jeremy, "This could never happen now."

"I can't believe that this is what a good time was like back then. How do we go back?" he said.

He took the words right out of my mouth. There we were, craving the old-time simplicity, the wholesomeness, and even some of the difficulties of a life lived in full and present engagement, without the assistance, speed, and conveniences of a tech-dominated world.

On the drive back to New York, Jeremy found some 1950s music on the radio, and we talked and sang the whole way home.

I felt so grateful. Especially when I discovered that some in this tech-dominated age hadn't been so lucky.

IN 2003, A COMPANY CALLED LINDEN LAB CREATED *SECOND Life*, one of the first virtual online worlds. As of 2017, it's still alive and well, with the company claiming about a million subscribers. Participants in *Second Life* create virtual selves, called avatars, who live in a fictional space where they can engage in activities, have careers, build businesses, share hobbies, play sports, fight crimes, be emotionally and sexually intimate, and participate in commerce, buying and selling services and products for virtual currency that can be converted into real money. The imagery you can create—

from the virtual home you live in, to the virtual restaurants you eat at, to the virtual clothes you wear—can be very visually alluring. There are mansions with swimming pools and gorgeous views that can be purchased and inhabited for the not-so-hefty price tag of (when translating the *Second Life* currency, called Linden dollars, to U.S. dollars) $16. There are high-end lines of furniture, jewelry, and sports cars for sale for a tiny fraction of what real ones would cost in the real world. It's a place where you can have relationships of all kinds with others who take on their own made-up, make-believe, virtual identities in this made-up, make-believe, virtual world. Complete with kids. Participants can create a whole virtual family together. And they do.

The *Second Life* tagline is "Your world. Your imagination."

But imagine this . . .

You wake up, get dressed, have coffee, go to work. Your spouse, let's say in this case your wife, stays home. While you're at the office, she goes into a room of her own, puts on headphones and a microphone, and turns on *Second Life*, immersing herself in a world where she has taken on the persona of a virtual version of herself, a cartoon she has created with whatever physical characteristics she wants. She talks as this avatar, moves about in different worlds, and eventually engages with another man's cartoon avatar, behind which is a real-life man wearing a microphone and headphones, locked away in a room of his own in an actual home somewhere. The characters they create, these avatars, can kiss, talk, have sex, build a virtual family, whatever they like, in this virtual world. Then your real-life wife takes off her microphone and headset, exits the room, and mindlessly waves hello as you arrive home. You have no idea about the virtual world she engaged in all day, or the real-life man behind the avatar she engaged with. Forget virtual cheating. This is a whole alternate, virtual existence.

Disturbing, right? I sure thought so.

In the documentary film *Life 2.0*, director Jason Spingarn-Koff features stories about people whose real lives have been upended by their compulsive participation in *Second Life*.

I was astonished by a woman who became a virtual designer, creating virtual clothing, skin care, and shoe lines for avatars (or for the people behind the avatars, I should say) to purchase in *Second Life*. It amazed me how she sat at her computer for fifteen to twenty hours a day, focusing her energy and creative talents in virtual stores rather than attempting to launch those very same enterprises in real life. The products she made were impressive, and she was a talented designer. People in *Second Life* were buying them. With purchases of $5 here, $1 there, she was making a real income of, as she put it, well into six figures.

I found myself shouting at the television: "Take that talent and step out into the real world!"

Also featured was a married woman who entered *Second Life* and met a virtual married man with whom she had a virtual emotional affair. The affair crossed over into Skype and turned physical when they met in person. They initially wound up sharing some real-life, temporary escape time that echoed the fantasy sentiment of their *Second Life* virtual relationship—no chores, no everyday challenges, just what seemed like mini-vacations together. Ultimately, they separated from their spouses and decided to live together for real, which is when things fell apart. They argued a lot, and she wound up feeling like she never knew him at all. Her last statements in the documentary included this: "I want to get across that I am a victim, and that doesn't mean I'm stupid. I believed he was really the person he presented, but that's a silly notion on my part because he's not real."

I saw it differently. She wasn't a victim. She went into that virtual world willingly, allowing herself to be drawn in without really questioning the dangers it could bring. That being said, I do believe that her realization that someone could appear a certain way in the virtual world but be nothing like that in the real-life one is a valuable insight for us all. These characters who exist in the virtual world are just that, characters. Who they really are—what they value, how they treat people, how they handle real-life issues—is often a big mystery.

Yet, these fictional characters can cause real-world problems. In a November 14, 2008, article in the *Telegraph* titled "Woman Divorces Husband for Having a 'Virtual' Affair on Second Life," writer Richard Edwards revealed the case of a woman who filed for divorce after she "discovered her husband's character having sex with another female player online." The woman said that her husband "committed adultery" with the animated woman and said that it was the second time she had caught his character cheating on her. More interesting? The lawyer who received the call said she "wasn't at all surprised." That was her second case—that week.

Virtual Adultery and Cyberspace Love is another documentary about *Second Life* that was made in 2008 and played on the BBC Two show *Wonderland*, a series that features short films about contemporary British life. The film focuses on two couples that meet and fall in love in *Second Life*. As the narrator explained, it's the ability to reinvent oneself, to look the way you want, to live the way you want, that is so appealing. As one woman interviewed said of her avatar, "She's everything I should have been . . . I was in control again . . . [I] got my life back."

In the main story of the documentary, there is a family in which a wife and mother of four started spending fourteen (you read that right, fourteen) hours a day on the computer. The husband noted, right at the beginning of the documentary, that he knew something was wrong when his wife pushed him out of the bedroom. She said it had nothing to do with *Second Life*—that she had been unhappy for a while. He then started sleeping in the living room, tried to keep life stable, and focused on helping the kids with homework and meals. The man talked about how he loved his wife and that she was just going through something. The kids were confused and angry. Some told their mother that they felt the computer was more important to her than they were. One said he thought their mother would be happier if they weren't even born, so she could do what she wanted.

As the woman explained, she'd been home for nine years with the kids and had become depressed, saying she would cry when doing laundry or the dishes. Looking for stimulation and escape, she made an avatar in *Second Life* and soon she was fully involved. Eventually, she met another avatar, opened up to him, and was spending most of her day with him in *Second Life*.

Then, similar to the woman in the *Life 2.0* documentary, she started to have an affair with him online through verbal descriptions of how they would touch each other and more. Things escalated. But the woman noted, "It's just human nature to want that human touch, that will just constantly eat at you . . . [I]t's like having that cookie sitting right there, but you can just get so far, you can almost taste it, you can see how good it is, but you can't, you can't get it, and that's tormenting."

The man with whom she was involved said, "You describe what you're doing to someone. It's a lot of imagination, yeah."

The woman: "Ninety-five percent of sex is in the brain, so you can make it something really incredible."

The man: "Only thing you can't do is touch, but I do miss, in any relationship, you know I do miss the sort of tactile part of it, I do miss the touch quite a lot."

Then the woman was pressured by the man to meet. She couldn't. She revealed, "My real-life situation wasn't such that I could just hop on a plane and go see him." They fought. It went on and on. He broke up with her. Again, this was all online, taking place in a cartoon world with cartoon characters. But the woman was, in real life, extremely sad about the breakup. She could hardly function. Her behavior affected and upset her real-life family.

Finally, the woman needed to know if she was interested in the person behind the avatar, the real man with whom she was having a virtual affair. She decided, despite the protests of her husband, to fly out and meet the man. It was incredibly strange to watch. During the time they shared together in the film—on a tour bus, on a picnic, even in his apartment on the couch—they barely made eye contact, looking outward as they spoke in halting, stilted sentences. They seemed like real-life strangers.

Eventually she went home to her ever-loyal husband—though, in the end, even the married couple's dialogue was more self-reflective reporting for the camera than it was a conversation with each other. He acknowledged that, no matter what, he would always be there for her and always love her. She appeared uncomfortable, bothered, and disengaged, wanting to escape, her eyes wandering the room. It was hard to tell if there was any relationship left at all.

Her attempt to substitute one world for another was like any other attempt to ditch reality via some kind of escape. As with those who use drugs or alcohol to gain entry into mind-numbing avoidance, her deep immersion in the virtual world was all-consuming and ultimately disastrous.

Second Life founder Philip Rosedale appears to believe that there's little difference between reality and virtually simulated reality. He said in *Life 2.0*, "Things are real because they're there with us, and we believe in them, and if they're simulated on a digital computer, versus sort of simulated by atoms and molecules, it doesn't make any difference to us."

I disagree. Virtual worlds have their limits. Humans are still human and crave real-life, human things. That's why many users of those virtual worlds seek to carry relationships over into real life, where touch, eye contact, and greater intimacy have a chance to grow in a human-to-human, person-to-person, screen-free way.

Life 2.0 and *Virtual Adultery and Cyberspace Love* were the first documentaries I ever watched where I felt the need to shower immediately afterward. I don't think I'm the only one who felt that way. Though it didn't bother Mr. Rosedale, who said in *Life 2.0*, "People inevitably, you know, approach it with a 'How could that be?' kind of a question. 'How can the virtual world be real?' But I think, in the years to come, we'll realize that the question was just more 'Why not?'"

I'll tell you why not. Because I'm not an avatar. I'm a person, not some wooden Pinocchio dancing on strings pulled by Silicon Valley. My relationships involve real talks with real people.

What happens to love tomorrow if people welcome *Second Life* and sex robots today?

#

THIS HEADLINE MADE ME SAD:

"The End of Young Love: Dating Is in Decline Among the 'i-Generation,' Study Finds"
(Telegraph.co.uk, Education, September 11, 2017)

The article reveals a study by a psychology professor who discovered that those born between 1995 and 2012, what the professor called the i-Generation, are "noticeably less interested in romance than their millennial predecessors." The professor, Jean Twenge of San Diego State University, found that "teenagers from this group have grown up with social media and smartphones, meaning they spend far more time socializing with one another online than they do in person."

I wanted to cry. But it's not just an "i-Generation" problem. It seems to me that too many people of all ages are losing their grip on romantic reality.

One of my favorite movies is *The Notebook*, based on the Nicholas Sparks novel. Set in the 1940s (reason #1 why I like it), the movie tells the story of Noah Calhoun and Allie Hamilton, who meet at a small-town carnival (reason #2). The couple is forced apart by Allie's parents, but Noah writes letters to Allie every single day for a year (reason #3), and I mean real, live, handwritten letters (reason #4). In later years, Allie grows sick, suffering from memory loss as a result of Alzheimer's, and Noah reads her a handwritten journal she once wrote to him (reason #5) about their life together to help her regain her memory.

Obviously, this movie hits all of my touchstones for romance.

Yet, it's a story that wouldn't happen today.

Few people write handwritten letters these days. If they lived in 2018, Noah and Allie wouldn't need to. The couple would be texting, or at least know each other's status and whereabouts via Instagram or Facebook. And the handwritten journal? Possible, but unlikely. More likely a software program on a computer, filled with computer files, folders, and typed text that would be missing the curves and angles of someone's personal touch.

I miss handwriting. Journals. Letters. I once read *I Love You, Ronnie: The Letters of Ronald Reagan to Nancy Reagan*, a compilation published in 2000 of the letters, many of which were handwritten, that President Reagan wrote to his wife throughout their marriage. It's a personal documentation of his life as an actor, governor, and president, revealing the true heart and mind of one of the most powerful leaders in history. There's something about Reagan's handwritten words that is just different. The loops of his letters make him seem present on the pages, revealing something that transcends time and invites you into the depth of his moments.

No email could ever do it quite that way. What happened to all of that special, thoughtful documentation and communication? Is it dying? Is it dead? Back in the day, if you wanted to tell someone you loved them for the first time, it took energy and courage and effort. You had to be inspired and moved, but most importantly, you had to be present. You had to show up in person or pick up the phone, dial, and get the words out somehow. Or maybe you'd take the time to pull out a nice piece of stationery or go to the store and select a postcard. You'd sit down at your desk, collect your thoughts, build up some courage, take hold of your pen, and write from the heart. Then you'd put it in an envelope, address it, stamp it, and bring it to the post office.

Thought. Detail. Effort.

Now? You can text an emoji of an eye and a heart and type a *U* while you're on the bus thinking about something else altogether. Are we better or worse off for that tech-assisted convenience? Much of the time we are worse off. Because it's the thought, time, and effort that strengthen real love, that give it its legs and wings, and that keep it strong through all of life's waves and thunderstorms.

# #

I DISCOVERED THE ANSWER TO MY QUESTION:
Who has time for love?

It was this: *Everyone.* If you feel like you don't have the time, make it. It's so incredibly worth it.

I believe in ROMANCE. I believe it's worth fighting for. My first few dates with Jeremy involved long walks and talks in beautiful settings—on the Manhattan High Line, over the Brooklyn Bridge, along the East River. Just us, no noise. No devices in sight.

We have kept that boundary in our relationship.

When we're together, unless we're doing work side by side, we put the devices away. When we go out for a walk, we leave the phones home. If we're together and decide we want to use our phone cameras, we put them on airplane mode to keep the other features quiet. The only thing on our dinner table is our food, and sometimes our elbows when we lean in to hear each other better or look at each other. For every holiday, and sometimes just because, we write each other handwritten letters.

We all have to find a way to remember what's worth fighting for, for the Ferris wheels, the sunsets, the roller coasters, and the romance. For our very humanity.

## CHAPTER 12

HUMANITY MUST WIN, OR WE ALL LOSE

Who's driving this technology train?

THE FLUX CAPACITOR IS CHARGED. THE PLUTONIUM IS ON board. The DeLorean is ready to go. Marty McFly steps in.

The Present Time on the dashboard reads October 26, 1985.

The Destination Time: October 26, 2017.

Marty buckles his seat belt, turns the key, braces himself for temporal displacement. The engine revs up, the car speeds ahead and hits eighty-eight miles per hour, and *zoom, flash*, it disappears, leaving nothing but a trail of parallel tire marks and foot-high orange flames.

A moment of nothingness, complete quiet, and then—

Swoosh. Whoosh. Screech.

Dr. Emmett Brown's time travel machine lands right back in the mall parking lot. The winged doors of the DeLorean rise up. Marty emerges, looks around, and sees rows of parked cars in the area up ahead, close to the mall entrance. He runs his hand through his hair, squints through the sunshine, adjusts his dark orange bubble down vest, and turns to bump into a person coming right toward him. The guy's head is down, looking intently at a small rectangle in his hand, and there are what appear to be tiny headphones in his ears.

"Excuse me," Marty says.

The person doesn't look up, continues on.

Marty walks to the mall's entrance and approaches the first store. The green sign reads "Starbucks." Thinking the store is named after the first mate of the ship *Pequod* in the novel *Moby-Dick*, Marty enters, expecting to find some books, boats, or fish. Instead, he sees a dozen or so people at various tables with paper cups nearby. Most wear headphones. Some are looking at small rectangles like the ones Marty has already seen, or at even bigger squares with typewriter keys attached. No one seems to notice him. Marty approaches one of the people.

"Excuse me, can you—"

She doesn't notice him, headphones in, rapidly typing on her small rectangle.

Fearing he's entered some strange world of robots, Marty goes to the counter. A cheerful worker wearing a green apron and a big smile says, "Hello!"

Finally. Relief. "What's going on here?" Marty asks.

"What do you mean?"

"You're the first person to look me in the eye. Everyone else is looking down at—"

Bzzzzz. The man behind the counter holds up a finger. He picks up his own rectangle, looks at it quickly, then puts it back down.

"What can I get you?"

Marty wonders what bizarre hold the rectangles have on these people. Are they some kind of magic? But before he can ask, someone behind him in line nudges him forward. "Let's go, buddy."

"Which coffee would you like?" the man behind the counter asks.

Marty looks up at the menu. "Does that say two dollars and seventy-five cents?"

"Yes."

"For coffee?"

"A tall, basic black, yes."

"Tall?"

"Small."

"Since when is tall small?" Marty asks.

"Do you want coffee or not?"

"Move it along, mister," says another man from the line. But then that man's rectangle buzzes and he is otherwise consumed.

"Now which coffee would you—"

Bzzzzz. The man behind the counter picks up his rectangle again, gives it a glance, then shoves it back in his pocket. "Sir, there's a line. What do you want?" he asks Marty again, a bit more agitated.

"I want someone who will talk to me."

"Next."

Marty travels on to a nearby park and meets some people who do talk to him. They explain that the tiny rectangles are phones. Odd. He never sees any of those people use the rectangles to actually make a phone call. His new acquaintances show him how the phones work, including a bunch of features that leave him confused, tired.

"How do you keep up with all that?" he asks them.

"What do you mean?"

"It seems constant. You have to look at stuff, check in on stuff, read stuff, acknowledge stuff. It's . . . a lot of stuff."

"That's true."

"Then how do you have time for anything else?"

His companions look at each other, then back at him. *Bzzzzz* yet again. One guy looks down at his phone, interrupting Marty's train of thought. The others then check theirs. This seems to occur every couple of minutes as one person or another stops to look down and cater to his or her rectangle.

The crew invites Marty to a party. He's hopeful at last. Some time to talk and get to know these people. He goes to the party, but amid the talking, everyone stops over and over again to look at their rectangles. They hold the devices up midsentence, focus them on an object (mostly themselves, it seems), click a button, and show Marty and each other the photo. He watches as they examine the images, critique them, and, as they explain, send them off into someplace in the air where a bunch of other people who aren't at the party can see them, too. They take a photo of Marty with them, but, no, someone's hair is messy. So they take another. No, the girl doesn't like the way her chin sticks out. Another. No, this one makes Marty look weird, they say. Finally, they get one they like. They all send it to each other and Marty waits while they type into their rectangles. They have to choose a "hashtag," one says, to capture the moment. They explain that this hashtag puts a bunch of connected words after a number sign, #, a symbol Marty has seen on typewriter keyboards. They go back and forth for a few minutes trying to settle on a hashtag they like. Then they need a filter. Marty asks why. "See how your eyes look crinkled here? Now watch it go away." To Marty, that doesn't sound like capturing the moment at all.

Marty doesn't bother to ask any more questions and they don't seem to notice when he leaves them and heads over to another group. A girl and her boyfriend are talking while another guy stares into his rectangle. Marty introduces himself. They are welcoming and nice. Suddenly, the girl's rectangle buzzes. She and her boyfriend look down and start laughing. Marty forgets what he was saying. He waits a few seconds.

"Hey, guys, my eyes are up here," Marty says, waving a hand in front of them.

They show him a photo of their friend's ex-boyfriend's cousin's new cat on someone's book of faces or something or other. Marty politely smiles. He excuses himself to grab a soda from the kitchen, taking a minute to look around. Rectangles everywhere. There are so many people not even looking at each other, he doesn't understand. What was the point of coming to the party at all?

By the end of the night, Marty, much like me very often, is thinking: *Take me back to 1985. PLEASE.*

THAT WAS MY MAKE-BELIEVE VERSION OF HOW MARTY MC- Fly might feel if he landed in tech-saturated 2017. I worry that he might think that humanity has already lost. But the good news is—it hasn't. If I could pull through the thickets of the tech jungle, we all could. There was one day in particular when I first realized that things had finally changed for me.

It was a small, simple moment, not a big, splashy one. However, it spoke volumes about how I had changed. It happened toward the end of my journey, as I was finishing writing this book, reflecting, reclaiming, and figuring things out for myself.

I was getting my hair done at my neighborhood salon. Trimmed, shaped, a clear gloss for some shine. I'd been sitting in the chair for a bit, talking with my stylist about the things one talks about with a stylist: life, love, travel, gathering her insights, gaining from her wisdom. We also ventured into the topic of which hair products she was using on me, as toxins and contaminants in our everyday products and food are the next creeping-normality Goliaths on my investigative agenda. I was asking a million questions. We were still in conversation when we moved over to the next step in the process, the washbasin. I sat down, put my head back, and let the relaxing flow of warm water stream through my hair.

Then there was a sudden panic around the basin. The water was running close to my ears, so at first I wasn't really sure what was going on. The stylist tapped me on the shoulder. I raised an eye in her general direction.

"It's your phone. It's vibrating over there."

"Oh, that's okay."

"I'll get it for you."

"No, it's okay, I'm fine."

"You sure? Are you okay?"

"Yeah, I don't need it right now. No worries." I put my head back in the basin under the warm water. It felt so good to relax in that moment, to have a minute to think about absolutely nothing.

But then I heard more panic, more rustling. Reluctantly, I lifted my head again and looked over at my stylist.

"It's vibrating again. I think someone's calling this time. Do you want me to get it for you?"

"It's okay, but thanks. Really, it's fine."

"Really?"

"Yeah, I can just check it later."

She paused for a few seconds. "I don't think that's ever happened," she said with a laugh.

"What?" I asked.

"I've never had a client hear their phone buzzing from out of reach and not want it right away. I think I'm in some kind of shock."

"Really?"

"Oh, yeah. If they don't already have it on their lap, in their hand, or plugged in right next to them, I have to run like a maniac to get it. And then, when they sit over there and I'm cutting their hair, they're pretty much on it all the time."

"Do they talk to you?"

"Yeah, they do, in between whatever they're typing. I hear lots of sentences that start and don't finish."

"That's sad," I said.

I quickly went through a whirlwind of emotions.

I felt bad for this poor stylist who had to be rushing around the salon every time someone's phone buzzed.

I felt sad about the partial sentences and distracted conversations people constantly threw her way.

I also realized something. *Uh-oh.* Those clients had been me. I had behaved in the same way in the past. Visuals flooded my mind of past trips to the salon, where I had embodied everything I hated hearing about in the stylist's story that day. Me checking my phone, me panicking at the basin when it would vibrate out of reach, me partially listening to her while I typed something to someone, somewhere, in the virtual ether.

Suddenly I couldn't stop smiling.

My tech-tic-have-to-check-it was *gone.* Here I was at the hair salon, relaxing and recharging. Not only wasn't I obsessively plugged into my device, but I wasn't even battling a desire to check it anymore. I had passed the stage where I had to resist the compulsion to immediately see or read whatever was coming in, or to instantly respond to whoever was trying to reach me. The compulsion was gone, and in its place was a feeling of contentment, a genuine comfort with the idea that whatever it was that was beckoning for my attention could wait. I imagined that this was what an addict felt when the withdrawal symptoms subsided. No panic, just peace. I was finally present.

VICTORY.

I had won.

I BELIEVE WE ALL WILL WIN.

Underneath all those hypnotic trances into iPhones or tablets that I see everywhere, every day, I can still see a human spirit that craves and values and will fight for a world where *people* are in charge, not some buzzing device in their pockets.

I know we will surface from the tech pollution in which we now swirl and find a way out of the Tech Takeover with our heads and hearts intact. I believe that we will ultimately release ourselves from our bondage to our devices and gadgets and lead our own lives again. I know that's true. It's always been true of us humans. When the transistor chips are down, we come through. All we need to do is get our heads back in the game and make our own thinking a priority.

If each person, one by one, takes notice and takes a stand against the thoughtless, mindless onslaught of technology, we will once again be humans in charge of our gadgets and our devices, in charge of our time and our lives. We will have learned how to use the gadgets instead of allowing them to use us. *We* will be in control.

We won't talk on our cell phones in public spaces and annoy others with our conversations on the bus, at the beauty salon, in the gym.

When we talk on the phone, we will focus on who it is we're talking to and not also be on our computers, our iPads, our social media, our who knows what. We will be present and engaged.

We will pick up the phone to say, "Good night, I love you" rather than texting it.

We will text and direct message for scheduling, planning, and basic communications, but not for the more serious stuff that deserves an accompanying tone of voice or facial expression.

Social media won't be used anonymously and cowardly to promote unkind and hateful agendas from one's hiding place behind a keyboard.

We will take time off from our devices. We'll keep our cell phones off the dinner table and out of the bedroom, and spend tech-free time with friends—time that is full of conversation, laughter, memories, and all that good stuff, free of distraction from buzzing devices.

We'll remember that just because all of our work sits right there in the laptop, and the laptop *can* be taken everywhere, doesn't mean that it *should* be taken everywhere. We will define a workday that makes sense for our lives and doesn't interfere with our nonwork priorities.

These are small things. But if we do them inch by inch, we will cause a momentum switch—a big one.

TWO OF MY HUSBAND'S FAVORITE BOOKS, AND ONES I'VE come to love as well, are Ayn Rand's *The Fountainhead* and *Atlas Shrugged*. This comes as no surprise, I'm sure, to anyone who knows my politics. I'm drawn to Rand's philosophy of individual empowerment, self-reliance, purpose-driven work, competition, opportunity, and the accountability of I, myself, to me. This has always resonated with me. The idea that if we manage ourselves—if we each work diligently to do the best that we can, without harming others—then it will be the best thing for each of us and all of us.

One weekend, while reviewing *The Fountainhead* in the state of mind of this book and my issues with technology, I was struck by another Rand thought: "Man cannot survive except through his mind. He comes on earth unarmed. His brain is his only weapon . . . from the wheel to the skyscraper, everything we are and everything we have comes from a single attribute of man—the function of his reasoning mind."

The power of human thought. No computer can ever compete with that. Why? Because our minds are coupled with heart and soul and depth, with empathy and drive and commitment, with loyalty and love. A computer, no matter how fancy, is no match for a mind filled with all that good stuff. But do most of us still know that? Do we care enough to preserve our minds, our thought processes, by limiting the power of the technology that seeks to control both?

We must. We read one scary revelation after another about websites manipulating us, we get upset for a few minutes or a few days, and then we move on. We hop right back onto Facebook, hypnotized, staring into the social media ether. Then another event. We get angry about that one. Soon, we forget. We tread on, into our mind-numbing lives of unconscious robotics.

We human beings created these machines; they didn't create us. Why are we giving them so much power? Why not forge a new path to keep what we need from technology, get rid of what we don't, and be in charge of our lives again?

Some will suggest that the government should regulate the problem of tech addiction away.

WRONG. What are we, in kindergarten? We are grown people, with agency and accountability and volition. Fast-food restaurants and junk food manufacturers are going to keep on making fast food and junk food. They should if they want to. They have every right to produce what they want to produce. But that doesn't mean you have to eat what they are making. Similarly, it is *you* who decides how many hours of the day you are on your phone. It is *you* who decides if you value a real life over a virtual one.

You are in charge of *you*.

Many of those who live healthy lifestyles are healthy because they eat quality food and work out. Many of those who are unhealthy enjoy their junk food and sedentary lives. The same idea goes for tech. Those who are addicted to tech are glued to their devices for most of the day. Those who are engaging in real life are managing the usage of their gadgets and are in control.

We have choices, people. We make choices. That's the beauty of being human.

#

THERE ARE SOME OTHER SPECIFIC THINGS I'M HOPEFUL we'll do to make technology only a part of our lives, not all of our lives.

Some new acronyms for our acronym-heavy world:

SOSS

Save Our Sacred Space

I mean the sacred space of our brains.

Thought. Thinking. We cannot give our minds, our ability to think for ourselves, away. Our brains are a vital, precious resource.

They are also terrifyingly vulnerable.

The brain is amazingly adaptable, reacting to new information around it. This neuroplasticity, as the scientists call it, is a great thing when the brain is adapting to good input, like love, nature, and healthy relationships. It's a negative when it is adapting to bad input, like drugs, trauma, and too much technology.

Our priority must be to PROTECT OUR BRAINS. They are everything to us.

How can we do this?

Take Time for Quiet Thought. We are stressed out because we have too much stuff incoming, inundating, flooding through. Life is not meant to be that way. It is not the workings of a sane world to have constant tech pollution when you are just trying to live. There has to be a space for quiet, for stillness. Let your mind wander and wonder as you close your eyes and just PAUSE. This is how we enable our minds to solve our problems, how we think about who we are and who we want to be, who we want to be with, and how we want to live. If we sit still long enough inside our minds, we may even allow the subconscious—that underlying, intangible, mysterious part of the brain from which many an artist believes imagination springs—to do its thing, to nourish and restore the brain with instinct and intuition, gut feelings, and resolution. As the American painter Andrew Wyeth said, "I do more painting when I'm not painting. It's in the subconscious."

It's probably been so long since you sat back, closed your eyes, and just let your mind breathe for any decent amount of time that you may not remember how to do it anymore. I know I didn't remember for a long time. Babies used to be left in their cribs to look up at colorful mobiles above them. Kids were given free rein to be quiet with themselves, outside, and stare off into rain puddles. Adults used to lie on their backs in the grass, gazing up at dancing leaves as the sun twinkled through the canopy of tree branches.

Have you tried this in a while? I have. I was amazed by how good it felt. My brain received a dose of happiness that was very different from the burst of full-on, machine-programmed-to-create-addiction shock load that comes from video games, virtual reality rides, and a constant stream of "likes" on social media, begging for stimulation, triggers, and replays. Lying on my back in the grass and looking up at the sky was something sweeter. It was a feeling I had forgotten. It was nourishing and sustaining. I felt revitalized, gratified, and recharged, not weighed down, anxious, or drained, as I often felt when inundated with tech.

Build Up Those Relationships You Value. Everyone is spread so thin these days, in large part due to the tech boom and the ballooning of social media. You have 942 Facebook "friends," many of whom you haven't had a meaningful conversation with for as long as you can remember, if ever. Many others you've never even met. You're following 650 people on Instagram, only a handful of whom you'd want to spend time with on a Saturday night. You have 1,200 LinkedIn connections, most of whom you don't know well enough to trust their advice regarding a job referral. You're everywhere and yet nowhere. You get direct-messaged by strangers, friend requests from people you barely knew years ago and didn't want to know. You feel the pressure to add that Facebook "friend," to follow more Instagram accounts so that they'll follow you, to build more contacts to boost your LinkedIn presence, to broaden that circle of superficial "connections" that aren't connected to you at all. That's what everyone else is doing.

You're logged on, flooded with photos of mostly strangers and details about their lives, inundated with people in virtual spaces whom you would never call in an emergency, wouldn't trust for advice, and would rather not spend leisure time with. You're adding, responding, scrolling, group messaging. The numbers are rising, and yet your sense of fulfillment is on the decline.

Instead, why not think about the people in your life who really matter—people who would have your back no matter what. Whom would you call if you got great news or bad news? Whom do you trust? Who would check up on you to see how you were doing when times were tough? Who makes you laugh and enhances your day? Whom do you miss when you don't see them and get excited to reconnect with when plans are made?

Give those people the time and energy you'd otherwise spend scrolling through or acknowledging a sea of strangers or near strangers online in the pages of your social media or apps. Build up those real, quality relationships. Let them grow as you grow. There are only twenty-four hours in a day. The chunk of time you give to all those "friends" takes away from time you could be giving to FRIENDS.

Cultivate Creativity, Learn Something New, Struggle a Bit. Allow your mind to be creative. Draw, sing, design a new product, invent a solution to a problem. Stoke your imagination to think about things you never thought about before. Allow your thoughts to run wild. Take a walk, feel the rhythm of your feet, listen to the sounds around you, see what ideas come to mind. Take some time to put pen, pencil, or crayon to paper and doodle. Sit down at a piano, pick up a guitar, take a dance lesson. Learn a new language. Try out a new sport or workout routine.

Mentally and physically engaging activities help our brains grow and rebuild. They send our neuroplasticity in the right direction. In his book *The Talent Code*, journalist Daniel Coyle notes that taking the time and making the effort to learn new skills, to deliberately focus on engaging in physical, tangible activities, can actually help us feed and nurture our brains and make them better. This isn't wishful thinking; this is science. Coyle studied the idea that when we engage in activities that require struggle and work—that is, when things *don't come easy*—we build and rebuild the myelin sheath, the fatty substance around nerve cells that allows our entire nervous system to function properly. This then "increases signal strength, speed, and accuracy" of our "movements and thoughts."

Your neurons are reacting to outside stimuli all the time, whether you like it or not, so why not stimulate your brain with quality input so that you can shape your brain into something healthy, strong, and resilient?

Get off the computer, the phone, your social media. Overuse of all of that is poisoning your neurons. Get out into creativity, imagination, physicality, friends, and nature. These will save the sacred spaces of you, your body, your brain, and your thoughts.

TBT: Throwback Time

We can combine the best of what technology has to offer—efficiency, safety, access—with the best from the past, and use that combination of past and present to build our future.

Things from the past like:

Handwritten Letters. As I mentioned, I love a handwritten note. The effort, the focus, the personalization of it all, the thinking about what you want to say and not mindlessly texting and letting autocorrect fill in the blanks. It also helps us—again, our brains—to physically write. In his November 18, 2017, opinion piece in the *New York Times*, "Our Love Affair with Digital Is Over," David Sax, who is also the author of the 2016 book *The Revenge of Analog: Real Things and Why They Matter*, commented: "The limits of analog, which were once seen as a disadvantage, are increasingly one of the benefits people are turning to as a counterweight to the easy manipulation of digital. Though a page of paper is limited by its physical size and the permanence of the ink that marks it, there is a powerful efficiency in that simplicity. The person holding the pen above that notebook page is free to write, doodle or scribble her idea however she wishes between those borders, without the restrictions or distractions imposed by software." Sax also noted, "Web designers at Google have been required to use pen and paper as a first step when brainstorming new projects for the past several years, because it leads to better ideas than those begun on a screen."

I'm going to write more letters. Perhaps I'll send a nice long note to someone I appreciate, thanking them for a good deed that wasn't forgotten. I know my neurons will fire all sorts of pleasure chemicals in the process. Jeremy sometimes finds handwritten notes from me around the house. We have a small chalkboard on the wall by the kitchen. It started as a fun way to keep a shopping list, but we leave cute messages for each other on there sometimes, too, and I love that.

Neighborhoods. I shared the story of how my grandparents met, stumbling across each other in this store and that one, here and there in Brooklyn. That's the extreme end of romanticizing the little shop around the corner. But it's the truth. It did happen, it does happen. It could happen more.

What if more local shops sponsored sidewalk sales and street fairs? What if our towns, neighborhoods, schools, places of worship, and community centers created reasons for their residents of all ages to get outside and spend time together more often?

Events and activities would help local shop owners become acquainted with customers. People could interact and engage better with city workers like police officers, librarians, the town maintenance and sanitation workers. If someone personally knows the people who are responsible for keeping the streets clean, don't you think they'd be less likely to litter? If police officers know the kids on the block by name, won't the kids be less likely to look for trouble?

We don't need a wireless connection to connect.

These are not grand, lofty, difficult ideas. The solutions to a better life are simple. All they require is the desire to exchange "connections" for connection, coupled with a little sincere thinking and focused follow-through.

Physical Objects. In the fall of 2017, I was watching a TV series on Netflix, *Stranger Things*, about a group of kids, their families, and the odd things happening in their small town in Indiana, circa 1983. The plot was intriguing, but I have to admit, that wasn't what initially grabbed my attention. What got me was the setting of the show, the places and the objects, the artistic and scenic design recreating life in 1983. I wanted to jump through my television and make a call on the wall phone, play a board game, hide in a fort in my room, eat some Halloween candy without worrying about what was in it or who gave it to me, hop on my bike and whirl around town with my friends, then throw the bike in the bike rack, unlocked, and enjoy my special brown-bag lunch, complete with a special sandwich made by my special mom on Wonder Bread. (And I'm gluten-free.) I envied the distraction-free zones and the simplicity of a life I remembered from my childhood. Don't get me wrong, I also really liked the show's plot. But there was something so alluring about the time and place they were trying to recapture, and the nostalgia-rich physical objects that brought that time back to life for viewers.

I started thinking more about tangible objects. Family history, traditions, and customs were often handed down through physical objects like art, silverware, jewelry boxes, candlesticks, watches, and homemade items like quilts and furniture. This passing down of items regarded as dear seems to occur less and less these days. But there is something special about holding a spoon my grandmother held in her hand. Or sitting in a chair my great-grandfather made, covered with a blanket my aunt knitted. I like a photo album with printed pictures. I can turn the pages and feel the textures at my fingertips. I prefer books—real books with real pages that I can smell and feel and have a tangible interaction with—with a binding I can grip in my hand. Even the landlines could use a comeback, so that we can hold the phone in one hand, cradle the colorful cord with the other, sit down in the spot where the phone is without walking around everywhere, look out the window, and focus on the call.

Consider the important objects in your lives. Cherish them. Enjoy them, and one day maybe pass them along to someone whose life could be enriched by them, too.

I even miss tangible money. Shiny coins and crisp dollar bills. Sure, they're still around, but how many people carry tangible money these days? A friend and I were talking about the app Venmo recently. It works much like other mobile payment apps, such as PayPal, Google Pay, and Chase QuickPay, but Venmo has a disturbing side feature. In its default setting, it allows each user to see which sources other users have received money from and sent money to, as well as the reason for the money transfer. This was not appealing to my friend, who went to pay her babysitter, only to see that the payment that preceded hers was from her babysitter's boyfriend, labeled "sexual favors." This she did not need to know. Venmo often becomes yet another look-at-me, look-what-I'm-buying-and-doing, I'm-important-because-you're-paying-attention-to-what-I'm-doing sad by-product of the tech boom. It seems to me that it would unfortunately also make a great tracking mechanism for stalkers.

These services are very convenient, sure, but at what life cost? Once again, people have stopped thinking before using. I'll stick to plain old cash and a credit card. I'll also keep a few gold coins in my drawer, just in case, just for fun, as mementos of times worth remembering.

FOMOS

Focus on Making Occasions Special

Forget the FOMO: fear of missing out. How about finding out what you're missing right there in front of you. If we are going to take the time to be with other people, let's make being with other people the priority of that time. That is the whole point.

How do you know if you're doing that? There are a few metrics by which I measure my ability to focus on the occasion at hand:

The Eye Test. If you have eye contact, you have engagement. Without it, you're disconnected.

What I find appealing: restaurants filled with people focused on the people who are at their tables, invested in the experience of being at that restaurant to enjoy that food in that space at that time.

What I find unappealing: restaurants where couples or groups of friends have their heads down, each on the phone, or where families are at a table and the kids are on their tablets, the parents on their phones. That energy is disengaging at their table, yes, but it negatively affects the feeling of the whole room, too.

What I find appealing: a party where everyone is fully present, participating in conversations or activities with each other.

What I find unappealing: walking into a party where everyone is engrossed in the rectangles in their hands.

What I find appealing: being at a performance, whether it's a concert at a huge venue or a middle school play, where everyone there is watching and listening and engaging with each other—and only doing that.

What I find unappealing: when I'm at a performance, and the audience is recording, checking their phones for something, taking a photo, sending it, posting it, FaceTiming their friend in the middle of it all. Awful. How about going back to the idea that—*if you're not there, YOU'RE NOT THERE.*

You're going to miss things. That's okay. You can't truly be in two or three or four places at once, so why try? It leaves you nowhere while ruining the experiences of those around you, who are left distracted by the distraction you created.

The Process of Process. I like to go through things step by step, to really experience them. I like a good narrative, a story, an arc, the growth that comes with each phase. I want to experience the beginning, the middle, and the end. I enjoy the process.

A friend of mine told me a story that nearly had me shouting from the rooftops (again). She and her younger sister were at a school play for her kids. Afterward, they were mingling with the cast, other parents, some of the families. The sister told my friend that she thought a nearby guy was cute. As it turned out, my friend knew him. The sister asked my friend for his name. Within minutes, the sister had followed him on Instagram, and he had followed her back. (All this while he was standing less than ten feet in front of her. They never spoke in person that night.) One thing led to another, and soon they were communicating often on social media, texting, going back and forth for about a week. They made plans by text to meet. They did. She, along with her friends, met him at a bar he was already at with his friends. That night, they spent the night together. They never saw each other again. Bing, bang, boom. Done. Neither of them thought that was odd.

Talk about cutting to the chase. I started thinking about how that might've gone down when I was twenty-two, the same age my friend's sister is now. Maybe I would've asked for a super-casual intro to him somehow, in person, at the play. We would've talked a little. Then maybe he would've asked for my number. A couple of days later, maybe he would've called my house and we would've talked on the phone for a while, then set a time to see a movie. Maybe we'd go to the movies, he'd drop me off at home, and then the next day he'd call again to see if I wanted to go out to dinner that weekend. And so on, and so on. We'd be getting to know each other, figuring out whether we liked each other or were even attracted to each other at all. The back-and-forth would've taken time and effort talking on the phone and in person.

In contrast, my friend's sister's situation was missing all the steps that made the whole thing worthwhile. Not only were they not stopping to smell the roses; they didn't even venture down the path or through the garden, and instead leapt right over the entire park to the exit gate. It was like listening to a story about two people who were living their lives in fast-forward. No bonds formed, no real effort, no quality time. Who knows what might've happened had they taken their time. They might've found that they liked each other, as real-life people, as friends, as lovers, as whatever they discovered.

That's not to say that one-night stands or purely physical interactions didn't exist before the tech boom. Sure they did. But they weren't so prevalent, and they weren't being fed by a ballooning app and social media system that encourages spreading yourself thin and taking easy ways out.

Don't you see? Silicon Valley needs you to behave in a way that keeps those apps thriving and that money flowing into their pockets. Investing in real-life people with real-life effort wouldn't sustain those apps or social media empires. It would starve them. So now, because so many have complied with the behavior those developers want, life is filled with too many stories where things rapidly go from start to finish, in a whiz-bang moment, often meaningless, disconnected, and unfulfilling, leaving behind feelings of vacancy, of nothingness.

We all need to remember that there's a certain joy and fulfillment about a process, about a journey and a discovery. If you're consistently doing things in a way that takes away all the meaning, then why bother doing them at all?

Shared Presence. It used to be that when people were together at an event, they would really be there. Present. Engaged. In the moment. Unfortunately, too many people don't even know what that means anymore. When they're at a birthday dinner or holiday party or wedding, they're constantly checking their cell phones, jumping back and forth between the people right there in front of them at the event and the people demanding their attention through cell phone texts, apps, emails, and more.

When phones get put away and heads stay up, people see things. They see each other—the quick, funny, sweet gesture, the subtle smile. They even see the eye roll, for heaven's sake. These moments make life rich, real, honest, and often rewarding.

When I'm having an event and want photographs, I hire a professional photographer, or barter with a friend-photographer, and then share those photos with friends. Same goes for video. I do what I can to have guests keep their phones in their cars, their bags, their pockets, or their coats. The phones don't receive an invite to the celebration; the people do.

There was a time when being present and engaged in others' company was the norm. I am doing my best to bring that energy back. Perhaps you will want to do so, too.

I'D UNCOVERED THE ANSWER TO MY QUESTION:
Who's driving this technology train?

It was this: Us. *We've been in the driver's seat the whole time, whether we knew it or not.*

After all, HUMANS MUST WIN. I'm not saying to throw away our tech. I still have a social media presence, an email address, and an iPhone. But now I know that we can, and we must, create boundaries, set some limits, and question new inventions, innovations, and the latest breakthroughs that may not be the greatest. Managing the technology in our lives is something we can all do. And we'll do it differently, each of us, because we each may have different visions for the kinds of lives we want for ourselves. The important thing isn't how we do it, or even what we do. What's important is that *we, the people*, are making those decisions. *We, the people*, are taking the time to think about the next big tech thing before we leap to buy it or embrace it. *We, the people*, are reflecting on our relationship to the tech in our lives, challenging it when necessary, and refusing to compromise our values and priorities for some new normal that doesn't feel normal at all. *We, the people*, are upholding healthy boundaries of tech communication when apps or social media seek to dismantle them.

What's important is that we're awake, questioning, and actively choosing the aspects of technology that enhance our lives while rejecting the ones that don't. I now use technology when I want to use it. When I don't, I stick my smartphone in my bag and party like it's 1999.

I genuinely believe that only good can come from living an empowered life where you control the role of Silicon Valley's ever-advancing tech overflow into your universe. Do it for yourself, your life, and the positive ripple effect it will have on your interactions and beyond.

I said it once and I'll say it again: I believe in people.

Let's create a world where humanity wins.

AFTERWORD: A LETTER TO THE NEXT GENERATION

Dear Next Generation,

When I graduated high school, I got a letter from one of my mom's best friends. It was handwritten, one of those old-fashioned letters on pretty stationery and everything. It was honest and funny, filled with some stories of how she met me when I was a kid and some of the silly things I'd done throughout my childhood that made her smile. She offered some advice she had learned the hard way and some hopes for my health and happiness in the next stage of growing up. The thing I remember most about it, though, was her tone. She didn't lecture me. She didn't talk to me like some kid who was about to make a bunch of mistakes, or should be told what to do, or even needed her words of wisdom. Instead, she wrote warmly but bluntly of a world that she felt was often rewarding, but sometimes deeply disappointing. She said that she hoped I'd find something helpful in her writing, whatever it might be—something that would help guide me as I journeyed through life. I don't know if I can do any of that for you, but I'm going to try.

I sometimes look around and think that it really must not be easy to be a young adult or a teenager these days. I know some older people say you don't have it hard, but don't listen to them. You have challenges before you that many of us didn't have to face at that age, not by a long shot. You have parents, teachers, acquaintances, social media contacts, and friends able to reach you on your phone or your tablet or your computer all day and night, and it's a lot to handle. You have to find room to get homework done and see friends and family face-to-face, and I imagine that your brain sometimes feels like it's on overload. With all the tech stuff coming into your devices all day long, twenty-four hours a day, I'm guessing it's hard to find time to even figure out who you are and what you like and love.

A lot of us didn't have to grow up with all of that going on and didn't have to juggle quite so proficiently. Because you grew up with a lot of that technology, you can manage electronics, multiple devices at once, much better than many of us can. Some of you will likely create tech inventions that will improve people's quality of life and, in some cases, save lives. Being tech-savvy has its perks, and you are going to usher in some products and devices that are sure to be amazing and life-changing in ways I probably can't even imagine.

Today there are apps, social media sites, and home devices you can talk to that will answer you back. Tomorrow there will be bigger inventions. Some will make our lives easier and richer. Others will make it harder and darker. The good news is that what you create, embrace, and reject is still up to you.

My hope is that when your creative juices flow and you build the next big tech thing, you pause to think about that device's potential effect on you, your life, your relationships, the values you care about, and the world at large, before it's released. Make sure that what you're about to contribute to the world feels good and positive to you. When faced with an app or feature or device that everyone is using, whose creators promise will make your life better, I hope you remember your power to say yes or no, to open or shut the door of your home to its knocking. I didn't always do that so well, and sometimes I found myself letting things into my life that weren't so good for me. Sometimes it was people with bad tech habits. Other times I just followed the trend of the latest and greatest invention without remembering who I was or what I cared about, or without even weighing whether that invention would nurture or injure the things that really mattered to me. I made those mistakes. I lost my way. I missed some important moments and saw some life-changing things I sometimes wish I could unsee. Because somewhere along the line, I forgot my ability to think, to question, and to do what was right for me. If there's a chance you can avoid that, that's what I hope for you.

Most of all, I hope you remember your power and your strength to build a healthy life no matter what Silicon Valley—or anyone else, for that matter—tosses your way. Every few months or so, I take a minute, even if it's just a single minute, to look at my life and the technology in it. Then the hard part comes of having the courage to question what's good or bad for me, of being honest with myself, and of getting rid of what inhibits my happiness.

I remember having a friend in high school who sometimes wasn't very nice to me. She would make me feel bad about myself. I struggled to cut the tie because I had known her for years and our families were close. One day, while enjoying a cherry Italian ice with my grandpa in the park, I told him that I felt like my friend wasn't a friend anymore and was making me unhappy. He reminded me that people only have the power to make you unhappy if you give them that power, if you give them a seat at the table of your life. He was right.

The same goes for technology, or anything else in our lives. The energy of your life is what you make it. So, when the next big tech invention comes along, do me one favor: Remember your mind and your heart. Let them be your guide.

Oh, and if you ever find yourself on a Ferris wheel at a carnival on a beautiful day, just as the sun is about to set, don't forget to look up.

With love and belief in you, me, and all of us together,

Jedediah

ACKNOWLEDGMENTS

I AM SO INCREDIBLY GRATEFUL TO EVERYONE WHO HELPED me throughout my journey of writing this book. You have each inspired me, challenged me, and supported me in ways I can't even begin to thank you for. But let me try . . .

Jeremy, my Bop-Bop. I loved living some of these stories with you. Thank you for listening, advising, and just plain being you. You have been my anchor through this incredible writing journey. Love you.

Mom and Dad. I am able to do what I do because I have two of the best people in the world supporting me in everything that makes me happy. Love you to pieces.

Caroline Sherman, my right-hand woman through this all. You helped me to discover my words when I got stuck, made me laugh like crazy when I needed it most, and were excited about my vision for this book from day one. Your wit, humor, and the way you really listened when I'd share a story, idea, or concern, and helped me figure out the best way to navigate it all, made me love this whole process so much. You're my favorite organizational genius. Thank you, thank you, thank you. Honored to now call you a friend.

Paul Fedorko, my literary agent. You've been compassionate, excited, warm, honest, supportive, and hilarious—all the things I needed you to be during the writing of this book. You embraced my idea right from the first meeting we had to discuss it, and you immediately got to work. You were there to help me make crucial decisions every step of the way. I couldn't have done it without you.

Eric Nelson, my executive editor. One of the first things you told me was that you wanted me to write a book I loved. Your guidance and insight have helped me to do just that. Thank you for challenging me even when I was stubborn—I'm so glad you did! Your commitment to my vision on everything from the concept to the tone to the cover helped me bring to life a finished product I'm so passionate about.

Eric Meyers, my associate editor. Thank you for your attention to detail and precision in moving this project along in such a timely, organized way. You provided much-needed observations that were such a pivotal part of this process.

Theresa Dooley, my Harper publicist. From our very first phone conversation, you were reassuring, efficient, and so motivated to get started. You have been an absolute joy to work with. Thank you for your passion for this project and go-getter energy right from the start.

Dina White and Matt Aversa, my outside publicity team. Your energy and dedication meant the world to me every step of the way. You helped to make this journey so exciting and fulfilling. I appreciate you both so much.

James Iacobelli, my art director. Thank you for building a book cover that really fits me and my vision. I know I was opinionated on this part (and drove you a little nuts, don't lie), but you did a phenomenal job and were so committed to producing a cover I would love. What a fun shoot day I'll never forget!

Jordan Matter, my cover photographer. You might just be as crazy as I am, which I love. When I signed this book deal, I looked at Jeremy and said, "We have to get Jordan." As always, you captured my energy perfectly, amid so much laughter and fun. My headshots, engagement party, wedding day, book cover—I can't wait to see what your camera captures next. Ten-minute challenge . . .

Vincenza Carovillano and Dora Smagler, thank you so much for working your magic on my makeup and hair for the cover. Most of all, do we ever stop laughing? I wouldn't have it any other way.

To everyone and everything I've crossed paths with in life who helped me to grow, learn, rediscover, and empower myself to ghost my cell phone to take back my life, I thank you.

SELECTED BIBLIOGRAPHY

Altman, Jamie. "The Whole Leslie Jones Twitter Feud, Explained." *USA Today College*, July 25, 2016, http://college.usatoday.com/2016/07/25/the-whole-leslie-jones
-twitter-feud-explained/.

Asano, Evan. "How Much Time Do People Spend on Social Media?" SocialMedia
Today.com, January 4, 2017, https://www.socialmediatoday.com/marketing/how
-much-time-do-people-spend-social-media-infographic.

Ayers, John W., PhD, MA; Eric C. Leas, MPH; Mark Dredze, PhD; et al. "Pokémon
GO—A New Distraction for Drivers and Pedestrians." JAMA (*Journal of the
American Medical Association*) Network, December 2016.

Bila, Jedediah. *Outnumbered: Chronicles of a Manhattan Conservative*. Self-published,
2011.

Booth, Robert. "Facebook Reveals News Feed Experiment to Control Emotions."
Guardian, June 29, 2014, https://www.theguardian.com/technology/2014/jun
/29/facebook-users-emotions-news-feeds.

Buxton, Madeline. "The Internet Problem We Don't Talk About Enough." Refinery29,
March 15, 2017, https://www.yahoo.com/lifestyle/internet-problem-dont-talk-enough
-194000785.html.

Camerota, Alisyn. "Why I'm Breaking Up with Twitter." Opinion column, CNN
.com, July 12, 2017, https://www.cnn.com/2017/07/11/opinions/dear-twitter-were
-done-camerota-opinion/index.html.

Carr, Nicholas. *Glass Cage: How Our Computers Are Changing Us*. New York: W.W.
Norton & Company, Inc., 2014.

————. "Is Google Making Us Stupid?" *Atlantic*, July/August 2008.

Cialdini, Robert. *Influence: The Psychology of Persuasion*. Rev. Ed. New York: Harper-
Business, 2006.

Cohen, Lisa. "New Report Finds Teens Feel Addicted to Their Phones, Causing Tension at Home." Common Sense Media, May 3, 2016.

Coyle, Daniel. *The Talent Code: Greatness Isn't Born. It's Grown. Here's How.* New
York: Bantam, 2009.

Danna, Tony. "Three Square Market Microchips Employees Company-Wide." PRLog,
press release distribution, River Falls, WI, July 2, 2017, https://www.prlog.org
/12653576-three-square-market-microchips-employees-company-wide.html.

Davidson, Darren. "Facebook Targets 'Insecure' Young People to Sell Ads." *Australian*, May 1, 2017, https://www.theaustralian.com.au/ . . . to-sell . . . /a89949ad016eee 7d7a61c3c30c909fa6.

Denizet-Lewis, Benoit. "Why Are More American Teenagers Than Ever Suffering from Severe Anxiety?" *New York Times Magazine*, October 11, 2017, https://www .nytimes.com/2017/10/11/magazine/why-are-more-american-teenagers-than -ever-suffering-from-severe-anxiety.html.

De-Sola Gutierrez, Jose, Fernando Rodriguez de Fonseca, and Gabriel Rubio. "Cell-Phone Addiction: A Review." *Frontiers in Psychiatry*, Frontiers Media S.A, October 24, 2016. U.S. National Library of Medicine, National Institutes of Health, Bethesda, MD. 201610.3389/fpsyt.2016.00175; https://www.ncbi.nlm.nih.gov /pmc/articles/PMC5076301/.

Duggan, Maeve. "Online Harassment 2017." Pewinternet.org, Pew Research Center, July 11, 2017, http://www.pewinternet.org/2017/07/11/online-harassment-2017/.

Dunne, Daisy. "Fall In Love with Harmony: World's Only 'Talking' Sex Doll Has 18 Personalities, Answers Your Questions and Even Remembers Your Favourite Meal." *Daily Mail* (online), April 3, 2017, http://www.dailymail.co.uk/sciencetech /article-4376310/Sex-doll-TALK-Robot-different-personalities.html.

Eagleman, David. *Incognito: The Secret Lives of the Brain*. New York: Vintage, 2011.

Edwards, Richard. "Woman Divorces Husband for Having a 'Virtual' Affair on Second Life." *Telegraph*, November 14, 2008.

"11 Facts About Cyber Bullying." DoSomething.org, 2014, https://www.dosome thing.org/us/facts/11-facts-about-cyber-bullying.

Evans, Dayna. "Please Don't Text Your Employees at 9 p.m." The Cut, July 17, 2017, https://www.thecut.com/2017/07/barstool-sports-ceo-erika-nardini-texts.html.

Foran, Clare. "The Rise of the Internet-Addiction Industry." *Atlantic*, November 5, 2015, https://www.theatlantic.com/technology/archive/2015/11/the-rise-of-the -internet-addiction-industry/414031/.

Goldsmith, Belinda. "Porn Passed Over as Web Users Become Social." Reuters, September 16, 2008, https://www.reuters.com/article/us-internet-book-life/porn-passed -over-as-web-users-become-social-author-idUSSP31943720080916.

Groth, Miles. *After Psychotherapy: Essays and Thoughts on Existential Therapy*. New York: You Can Help Publishing, 2017.

Hamilton, Isobel. "A Sex Robot Appears on a British Morning TV Show and It's as Uncomfortable as You'd Expect." Mashable, September 13, 2017, https:// mashable.com/2017/09/13/sex-robot-samantha-goes-on-this-morning/#oMQY YJOYomqV.

Harris, Michael. *End of Absence: Reclaiming What We've Lost in a World of Constant Connection*. New York: Current, 2014.

Harrison, George. "Hackers Could Program Sex Robots to Kill." *New York Post*, September 11, 2017, https://nypost.com/2017/09/11/hackers-could-program-sex -robots-to-kill/. Originally published by the *Sun*, September 9, 2017.

Hopson, John. "Behavioral Game Design." Gamasutra.com, April 27, 2001, https:// www.gamasutra.com/view/feature/131494/behavioral_game_design.php.

Hotz, Robert Lee. "Can Handwriting Make You Smarter?" *Wall Street Journal*, April 4, 2016, https://www.wsj.com/articles/can-handwriting-make-you-smarter-1459784659.

"How Much Time Do People Spend on Their Mobile Phones in 2017?" HackerNoon .com, May 9, 2017, https://hackernoon.com/how-much-time-do-people-spend-on -their-mobile-phones-in-2017-e5f90a0b10a6.

Ilardi, Stephen. "Why Personal Tech Is Depressing." *Wall Street Journal*, October 26, 2017, https://www.wsj.com/articles/why-personal-tech-is-depressing-1509026300.

"I Mug You, Pikachu!" *Economist*, July 16, 2016, https://www.economist.com/busi ness/2016/07/16/i-mug-you-pikachu.

"Insufficient Sleep Is a Public Health Problem." *Centers for Disease Control and Prevention*, 3 (September 2015), http://www.cdc.gov/features/dssleep/.

Itzkoff, Dave. "'Marvin's Room' Moves to Broadway with Women Front and Center." *New York Times*, June 22, 2017, https://www.nytimes.com/2017/06/22/theater /marvins-room-moves-to-broadway-with-women-front-and-center.html.

Kelly. "Why Do We Use the Term Cellular Phone Instead of Mobile Phone?" Gizmodo .com, September, 16, 2011, https://gizmodo.com/5840939/why-do-we-use-the -term-cellular-phone-instead-of-mobile-phone.

Kerner, Ian. "What Counts as 'Cheating' in the Digital Age?" CNN.com, May 16, 2017, https://www.cnn.com/2017/05/16/health/cheating-internet-sex-kerner/index .html.

Kim, Joseph. Gamemakers.com, March 16, 2014, http://www.gamemakers.com/the -compulsion-loop-explained/.

Lenhart, Amanda, Michele Ybarra, Kathryn Zickuhr, and Myeshia Price-Feeney. "Online Harassment, Digital Abuse, and Cyberstalking in America." DataSociety.net, Data & Society Research Institute's report, November 21, 2016, https://datasociety .net/output/online-harassment-digital-abuse-cyberstalking/.

Levin, Paige. "Teen Electrocuted After Playing on Phone in Bathtub." CNN.com, July 18, 2017, https://www.cnn.com/2017/07/18/health/teen-bathtub-electro cuted-text-trnd/index.html.

Lewis, Paul. "'Our Minds Can Be Hijacked': The Tech Insiders Who Fear a Smartphone Dystopia." *Guardian*, October 6, 2017, https://www.theguardian.com /technology/2017/oct/05/smartphone-addiction-silicon-valley-dystopia.

Louv, Richard. *Last Child in the Wood: Saving Our Children from Nature Deficit Disorder*. Chapel Hill, NC: Algonquin Books, 2005.

Martin, Sean. "Sex Robot Shocker: Almost HALF of All Men Will Use Erotic Robot Playthings, Says Survey." *Express*, December 19, 2016.

Maslow, Abraham. "A Theory of Human Motivation." *Psychological Review*, 50 (1943).

McSpadden, Kevin. "You Now Have a Shorter Attention Span Than a Goldfish." Time.com, May 14, 2015, http://time.com/3858309/attention-spans-goldfish/.

Meyer, Robinson. "Everything We Know About Facebook's Secret Mood Manipulation Experiment." *Atlantic*, June 28, 2014.

Mischel, Walter. *The Marshmallow Test: Mastering Self-Control*. New York: Little, Brown and Company, 2014.

Morris, Chris. "This Selfie Cost $200,000." Fortune.com, July 14, 2017, http://fortune .com/2017/07/14/selfie-cost-200000-art-gallery/.

Netflix. Letter to shareholders, January 2016, https://ir.netflix.com/static-files/0d8af 1e9-a638-422f-900b-20fdb8b2797e.

Notopoulos, Katie. "This App Lets You Find People on Tinder Who Look Like Celebrities." Buzzfeed, June 20, 2017, https://www.buzzfeed.com/katienotopoulos /this-app-lets-you-find-people-on-tinder-who-look-like?utm_term=.nm2n7dX2 2#.gjA5OGvll.

Powers, William. *Hamlet's BlackBerry: Building a Good Life in the Digital Age*. New York: HarperCollins Publishers, 2010.

Rand, Ayn. *The Fountainhead*. New York: Bobbs Merrill, 1943.

———. *Atlas Shrugged*. New York: Random House, 1957.

Reagan, Nancy, and Ronald Reagan, *I Love You, Ronnie: The Letters of Ronald Reagan to Nancy Reagan*. New York: Random House, 2000.

Rhodan, Maya. "U.K. Kids Spend Less Time Outside Than Prison Inmates, Study Says." Time.com, March 25, 2016, http://time.com/4272459/u-k-kids-spend-less -time-outside-than-prison-inmates-study-says/.

Rock, Margaret. "A Nation of Kids with Gadgets and ADHD." Techland, Health and Science, Time.com, July 8, 2013, http://techland.time.com/2013/07/08/a -nation-of-kids-with-gadgets-and-adhd/.

Sax, David. "Our Love Affair with Digital Is Over." *New York Times*, November 18, 2017, https://www.nytimes.com/2017/11/18/opinion/sunday/internet-digital-technology -return-to-analog.html.

———. *The Revenge of Analog: Real Things and Why They Matter*. New York: Public Affairs, 2016.

Schmidt, Michael S. "Boston Red Sox Used Apple Watches to Steal Signs Against Yankees." *New York Times*, September 5, 2017, https://www.nytimes.com/2017 /09/05/sports/baseball/boston-red-sox-stealing-signs-yankees.html.

Skinner, B. F., and C. B. Ferster. *Schedules of Reinforcement*. New York: Appleton-Century-Crofts, 1957.

Smith, Aaron. "Record Shares of Americans Now Own Smartphones, Have Home Broadband." Fact Tank, Pew Research Center, Washington, D.C., January 12, 2017, http://www.pewresearch.org/fact-tank/2017/01/12/evolution-of-technology/.

Spear, Jane. "Existential Psychology—History of the Movement." *JRank Psychology Encyclopedia*, http://psychology.jrank.org/pages/229/Existential-Psychology.html.

Sulleyman, Aatif. "'Killer Robots' That Can Decide Whether People Live or Die Must Be Banned, Warn Hundreds of Experts." *Independent*, November 7, 2017, https://www.independent.co.uk/life-style/gadgets-and-tech/news/killer-robots -ban-artificial-intelligence-ai-open-letter-justin-trudeau-canada-malcolm-turnbull -a8041811.html.

Turkle, Sherry. *Alone Together: Why We Expect More from Technology and Less from Each Other*. New York: Basic Books, 2011.

Turner, Camilla. "The End of Young Love: Dating Is in Decline among the 'i-Generation' Study Finds." Telegraph.co.uk, September 11, 2017, https://www .telegraph.co.uk/education/2017/09/11/end-young-love-dating-decline-among-i -generation-study-finds/.

Ward, Adrian F., Kristen Duke, Ayelet Gneezy, and Maarten W. Bos. "Brain Drain: The Mere Presence of One's Own Smartphone Reduces Available Cognitive Capacity." *Journal of the Association for Consumer Research*, University of Chicago, April 3, 2017, https://www.journals.uchicago.edu/doi/10.1086/691462.

Warner, Claire. "How Does 'Pokemon Go' Work? Here's Everything We Know About the Tech Behind the Augmented Reality Fad." Bustle, July 13, 2016.

Watson, Kaitlyn, and David C. Slawson. "Social Media Use and Mood Disorders: When Is It Time to Unplug?" *American Family Physician*, October 15, 2017.

Whipp, Glenn. "Amy Schumer Stays Honest, Authentic—and Doesn't Care What Anyone Thinks." *Los Angeles Times*, August 10, 2016, https://www.mercurynews .com/2016/08/15/amy-schumer-stays-honest-authentic-and-doesnt-care-what -anyone-thinks/.

Worley, Will. "Mother of Cyber Bullying Victim Pens Heartbreaking Open Letter In Response to His Suicide." *Independent*, October 2016, https://www.independent .co.uk/news/uk/home-news/mother-open-letter-cyber-bullying-victim-suicide -online-social-media-a7347531.html.

ABOUT THE AUTHOR

JEDEDIAH BILA is a two-time Emmy-nominated television host. She was cohost of the historic season twenty of ABC's *The View* and hosted the Lifetime special *Abby Tells All*. Prior to joining *The View*, Bila regularly cohosted Fox News' *Outnumbered* and *The Five*, and was a contributor on a wide range of Fox News and Fox Business programming. She has a master's degree from Columbia University and is a former high school academic dean and teacher. She has taught at the middle school, high school, and college levels. Bila currently resides in New York City.